3-35

Tables, Data and Formulae
for
MATHEMATICIANS

Tables, Data and Formulae
for
MATHEMATICIANS

compiled by

A. Greer, C.Eng., M.R.Ae.S.

Senior Lecturer in Engineering,
City of Gloucester College of Technology
and

D. J. Hancox, B. Sc., F.I.M.A.

Head of Department of Mathematics,
Coventry Technical College

Stanley Thornes (Publishers) Ltd

Acknowledgements:
The authors and publishers gratefully acknowledge permission to reproduce copyright material in the form of the table of logarithms, antilogarithms, natural logarithms, reciprocals and normal distribution, as originally compiled by Messrs. J. White, A. Yeats, and G. Skipworth for TABLES FOR STATISTICIANS published also by Stanley Thornes (Publishers) Ltd.

From BIOMETRIKA TABLES FOR STATISTICIANS, Vol. I Edited by E. S. Pearson and H. O. Hartley the following: Percentage points of t-distribution with incidental text by J. White; Percentage points of the χ^2 distribution with incidental text by J. White; the Control chart factors for the sample range based on the standard deviation with incidental text by J. White. Also from the same source the Distribution of the Kendall rank correlation for which additional thanks to Dr. M. G. Kendall are recorded.

The Institute of Mathematical Statistics publishers of THE ANNALS OF STATISTICS for their permission to reproduce the Critical values of r_s the Spearman rank correlation coefficient.

First published by Stanley Thornes (Publishers) Ltd.
EDUCA House, 32 Malmesbury Road, Kingsditch Estate
CHELTENHAM England

ISBN 0 85950 037 3

Typeset by J. W. Arrowsmith, Bristol
Printed in Great Britain by offset lithography by
Billing & Sons Ltd, Guildford, London and Worcester

Contents

Mathematical signs and abbreviations

Symbol	Term	Symbol	Term
[{( )}]	brackets	$\lim y$	limit of y
$+$	plus	$\to a$	approaches a
$-$	minus	∞	infinity
$\pm$	plus or minus	Σ	sum of
$\lvert a-b \rvert$	modulus of difference between a and b	Π	product of
		$\sqrt{x}, x^{\frac{1}{2}}$	square root of x
$\times$ or $\cdot$	multiplied by	$x^{\frac{1}{3}}$	cube root of x
$\div$ or $/$	divided by	e	base of natural logarithms
$=$	is equal to	$\log_a x$	logarithm to the base a
$\neq$	is not equal to	$\ln x, \log_e x$	natural logarithm of x
$\equiv$	is identical with	$\lg x, \log_{10} x$	common logarithm of x
$\triangleq$	corresponds to	antilog	antilogarithm
$\approx$	is approximately equal to	$\exp x, \text{e}^x$	exponential function of x
$\sim$	is asymptotically equal to	$n!$	factorial n
$\propto$	varies directly as	$\binom{n}{p}, {}^nC_p$	binomial coefficient
$>$	is greater than		
$<$	is less than	Δ, δ	increment or finite difference operator
$\geqslant$	is equal to or greater than		
$\leqslant$	is equal to or less than	D	operator $\dfrac{\text{d}}{\text{d}t}$
$\gg$	is much greater than		
$\ll$	is much less than	$\int y \, \text{d}x$	indefinite integral
i, j	complex number $i = j = \sqrt{-1}$	$\int_a^b y \, \text{d}x$	integral between the limits of a and b
$\lvert z \rvert$	modulus of z		
$\arg z$	argument of z	$\oint y \, \text{d}x$	around a closer contour
x_i	ith value of the variate x	σ	standard deviation of a distributed variate
$\bar{x}$	average of several values of x		
ρ	correlation coefficient	s	standard deviation for a sample
r	correlation coefficient for a sample	n	number in a sample
p	probability	w	range
$\angle$	angle	$\therefore$	therefore
$\subset$	one subset	$\triangle$	triangle
$\mathscr{E}$	universal set	$\cup$	union
A^{-1}	inverse of the matrix A	$\cap$	intersection
$\parallel$	parallel to	A^T	transpose of the matrix A
$\lvert$	perpendicular to		

Multiples and submultiples

Multiplying factor	Prefix	Symbol	Multiplying factor	Prefix	Symbol
10^{12}	tera	T	10^{-6}	micro	μ
10^{9}	giga	G	10^{-9}	nano	n
10^{6}	mega	M	10^{-12}	pico	p
10^{3}	kilo	k	10^{-15}	femto	f
10^{-3}	milli	m	10^{-18}	atto	a

Greek letters

A α alpha	E ϵ epsilon	I ι iota	N ν nu	P ρ rho	Φ ϕ phi
B β beta	Z ζ zeta	K κ kappa	Ξ ξ xi	Σ σ sigma	X χ chi
Γ γ gamma	H η eta	Λ λ lambda	O o omicron	T τ tau	Ψ ψ psi
Δ δ delta	Θ θ theta	M μ mu	Π π pi	Y υ upsilon	Ω ω omega

Chemical symbols and atomic weights

Element	Symbol	Atomic number	Atomic weight	Element	Symbol	Atomic number	Atomic weight
Actinium	Ac	89	(227)	Molybdenum	Mo	42	95.9
Aluminium	Al	13	26.9815	Neodymium	Nd	60	144.2
Americium	Am	95	(243)	Neon	Ne	10	20.179
Antimony	Sb	51	124.7	Neptunium	Np	93	237.0482
Argon	A	18	39.948	Nickel	Ni	28	58.7
Arsenic	As	33	74.9216	Niobium	Nb	41	92.9064
Astatine	At	85	~210	Nitrogen	N	7	14.0067
Barium	Ba	56	137.3	Nobelium	No	102	(254)
Berkelium	Bk	97	(247)	Osmium	Os	76	190.2
Beryllium	Be	4	9.01218	Oxygen	O	8	15.999
Bismuth	Bi	83	208.9806				
Boron	B	5	10.81	Palladium	Pd	46	106.4
Bromine	Br	35	79.904	Phosphorus	P	15	30.9738
Cadmium	Cd	48	112.40	Platinum	Pt	78	195.0
Californium	Cf	98	(251)	Plutonium	Pu	94	(244)
Calcium	Ca	20	40.08	Potassium	K	19	29.102
Carbon	C	6	12.011	Praseodymium	Pr	59	140.907
Cerium	Ce	58	140.12	Protoactinium	Pa	91	231.0359
Cesium	Cs	55	132.9055	Polonium	Po	84	(210)
Chlorine	Cl	17	35.453	Promethium	Pm	61	(145)
Chromium	Cr	24	51.996	Radium	Ra	88	226.0254
Cobalt	Co	27	58.9332	Radon	Rn	86	(~222)
Copper	Cu	29	63.546	Rhenium	Re	75	186.2
Curium	Cm	96	(247)	Rhodium	Rh	45	102.9055
Dysprobium	Dy	66	162.50	Rubidium	Rb	37	85.467
Erbium	Er	68	167.26	Ruthenium	Ru	44	101.9
Europium	Eu	63	151.96	Samarium	Sm	62	150.4
Fermium	Fm	100	(257)	Scandium	Sc	21	44.9559
Fluorine	F	9	18.9984	Selenium	Se	34	78.96
Gadolinium	Gd	64	157.2	Silicon	Si	14	28.086
Gallium	Ga	31	69.72	Silver	Ag	47	107.868
Germanium	Ge	32	72.59	Sodium	Na	11	22.9898
Gold	Au	79	196.9665	Strontium	Sr	38	87.62
Hafnium	Hf	72	178.49	Sulphur	S	16	32.06
Helium	He	2	4.00260	Tantalum	Ta	73	180.947
Holmium	Ho	67	164.9303	Technetium	Tc	43	98.9062
Hydrogen	H	1	1.0080	Tellurium	Te	52	127.60
Indium	In	49	114.82	Terbium	Tb	65	158.9254
Iodine	I	53	126.9045	Thallium	Tl	81	204.37
Iridium	Ir	77	193.2	Thorium	Th	90	232.0381
Iron	Fe	26	55.84	Thulium	Tm	69	168.9342
Krypton	Kr	36	83.86	Tin	Sn	50	118.6
Lanthanum	La	57	138.905	Titanium	Ti	22	49.9
Lawrencium	Lr	103	(257)	Tungsten	W	74	183.8
Lead	Pb	82	207.2	Uranium	U	92	238.029
Lithium	Li	3	6.941	Vanadium	V	23	50.941
Lutetium	Lu	71	174.97	Xenon	Xe	54	131.30
Magnesium	Mg	12	24.305	Ytterbium	Yb	70	173.0
Manganese	Mn	25	54.9380	Yttrium	Y	39	88.9059
Mendelevium	Md	101	(256)	Zinc	Zn	30	65.3
Mercury	Hg	80	200.5	Zirconium	Zr	40	91.22

Arithmetic

Sequence of arithmetic operations

When numbers are combined in a series of arithmetical operations:
  (i) First work out brackets.
  (ii) Multiply and/or divide.
  (iii) Add and/or subtract.
Thus $12+4+7\times(3+2)-8 = 12+4+7\times5-8 = 12+4+35-8 = 43$.

Ratio

A ratio may be expressed as a fraction and vice versa. Thus $2:3$ is the same as $\frac{2}{3}$ and $\frac{3}{4}$ is the same as $3:4$.

Percentages

To convert a fraction or a decimal into a percentage, multiply it by 100. Thus $\frac{2}{5}=\frac{2}{5}\times100\% = 40\%$ and $0.63 = 0.63\times100\% = 63\%$.
A percentage can be converted into a fraction or decimal by dividing it by 100. Thus $85\% = 85 + 100 = 0.85$.

Percentage profit

$$\text{Profit \%} = \frac{\text{selling price} - \text{cost price}}{\text{cost price}}\times 100.$$

Profit % is often referred to as *mark up*.

Percentage loss

$$\text{Loss \%} = \frac{\text{cost price} - \text{selling price}}{\text{cost price}}\times 100.$$

Margin

$$\text{Margin \%} = \frac{\text{selling price} - \text{cost price}}{\text{selling price}}\times 100.$$

Gross profit

Gross profit = turnover − cost price.

Net profit

Net profit = gross profit − overheads.

Discount

$$\text{Cash price} = \frac{\text{marked price} \times (100 - \text{discount \%})}{100}.$$

Commission

$$\text{Amount of commission} = \text{amount of sales} \times \frac{\text{commission \%}}{100}.$$

Value added tax

$$\text{Amount of VAT} = \text{price of goods} \times \frac{\text{rate of VAT \%}}{100}.$$

Rates

Rates payable per annum = rateable value of the property × rate in the £1.

Income tax

Taxable income = gross income − allowances.

$$\text{Tax payable} = \text{taxable income} \times \frac{\text{rate of tax \%}}{100}.$$

Simple interest

$$I = \frac{PRT}{100}$$

$$A = \left(1 + \frac{TR}{100}\right)P$$

P = principal (i.e. amount of money invested or borrowed)
R = rate per cent per annum
T = time in years
A = amount after T years

Compound interest

$$A = P\left(1 + \frac{R}{100}\right)^T$$

A = amount of the investment after T years
P = principal or amount invested
R = rate per cent per annum

If the interest is added n times per annum:

$$A = P\left(1 + \frac{R}{100p}\right)^{nT}$$

Depreciation

$$A = P\left(1 - \frac{R}{100}\right)^T$$

A = book value after T years
P = initial cost of asset
R = rate of depreciation (% per annum)

Shares

Amount paid for the shares = number of shares bought × price paid per share.

Nominal value of shares bought = number of shares bought × nominal value per share.

Amount of dividend per share = nominal value per share × $\dfrac{\text{declared dividend \%}}{100}$.

Stock

Interest payable on stock = nominal amount of stock bought × $\dfrac{\text{interest \%}}{100}$.

Bankruptcy

Declared dividend = $\dfrac{\text{net assets} - \text{amount owed to secured creditors}}{\text{total liabilities}}$.

Algebra

Factors

$(a + b)^2 = a^2 + 2ab + b^2$
$(a - b)^2 = a^2 - 2ab + b^2$
$a^3 + b^3 = (a + b)(a^2 - ab + b^2)$
$a^3 - b^3 = (a - b)(a^2 + ab + b^2)$
$a^2 - b^2 = (a + b)(a - b)$
$(a + b)^3 = a^3 + 3a^2b + 3ab^2 + b^3$
$(a - b)^3 = a^3 - 3a^2b + 3ab^2 - b^3$
$(a + b + c + \ldots)^2 = a^2 + b^2 + c^2 + \ldots + 2a(b + c + \ldots)$
$\qquad\qquad\qquad + 2b(c + \ldots) + \ldots$

4

Indices

$$a^m \times a^n = a^{m+n}$$

$$a^m \div a^n = a^{m-n}$$

$$(a^m)^n = a^{mn}$$

$$\sqrt[n]{a^m} = a^{m/n}$$

$$\frac{1}{a^n} = a^{-n}$$

$$a^0 = 1$$

$$(a^n b^m)^p = a^{np} b^{mp}$$

$$\left(\frac{a}{b}\right)^n = \frac{a^n}{b^n}$$

$$\sqrt[n]{ab} = \sqrt[n]{a} \times \sqrt[n]{b}$$

$$\sqrt[n]{\frac{a}{b}} = \frac{\sqrt[n]{a}}{\sqrt[n]{b}}$$

Logarithms

If $N = a^x$ then $\log_a N = x$ and $N = a^{\log_a N}$.

$$\log_a N = \frac{\log_b N}{\log_b a}$$

$$\log(ab) = \log a + \log b$$

$$\log\left(\frac{a}{b}\right) = \log a - \log b$$

$$\log a^n = n \log a$$

$$\log \sqrt[n]{a} = \frac{1}{n} \log a$$

$$\log_a 1 = 0$$

$$\log_e N = 2.3026 \log_{10} N$$

Quadratic equations

If $ax^2 + bx + c = 0$

$$x = \frac{-b \pm \sqrt{b^2 - 4ac}}{2a}.$$

If $b^2 - 4ac > 0$ the equation $ax^2 + bx + c = 0$ yields two real and different roots.

If $b^2 - 4ac = 0$ the equation $ax^2 + bx + c = 0$ yields coincident roots.

If $b^2 - 4ac < 0$ the equation $ax^2 + bx + c = 0$ has complex roots.

If α and β are the roots of the equation $ax^2 + bx + c = 0$ then

$$\text{sum of the roots} = \alpha + \beta = -\frac{b}{a} \qquad \text{product of the roots} = \alpha\beta = \frac{c}{a}.$$

The equation whose roots are α and β is $x^2 - (\alpha + \beta)x + \alpha\beta = 0$.

Any quadratic function $ax^2 + bx + c$ can be expressed in the form $p(x+q)^2 + r$ or $r - p(x+q)^2$, where r, p and q are all constants.

The function $ax^2 + bx + c$ will have a maximum value if a is negative and a minimum value if a is positive.

If $ax^2 + bx + c = p(x+q)^2 + r = 0$ the minimum value of the function occurs when $(x+q) = 0$ and its value is r.

If $ax^2 + bx + c = r - p(x+q)^2$ the maximum value of the function occurs when $(x+q) = 0$ and its value is r.

Arithmetic progression

The general expression for a series in Arithmetical Progression (A.P.) is a, $a + d$, $a + 2d$, $a + 3d$, ... $a + (n-1)d$.

The last term of the series is $l = a + (n-1)d$.

The sum of n terms in A.P. is

$$S_n = \frac{n}{2}(a + l) = \frac{n}{2}[2a + (n-1)d]$$

Arithmetic mean

The arithmetic mean of three numbers a, b and c which are in A.P. is

$$b = \frac{a + c}{2}$$

Geometrical progression

The general expression for a series in Geometrical Progression (G.P.) is a, ar, ar^2, ar^3, ..., ar^{n-1}.

The sum of n terms in G.P. is

$$S_n = \frac{a(r^n - 1)}{r - 1} \quad \text{use if } r > 1$$

$$S_n = \frac{a(1 - r^n)}{1 - r} \quad \text{use if } r < 1$$

If n is infinity and $r^2 < 1$,

$$S_\infty = \frac{a}{1 - r}$$

Geometric mean   If a, b and c are three numbers in G.P. then $b = \pm\sqrt{ac}$.

Permutations and combinations

The number of permutations of n different objects taken r at a time is
$$^nP_r = n(n-1)(n-2) \ldots \text{to } r \text{ factors}$$
$$= n(n-1)(n-2) \ldots (n-r+1)$$

The number of combinations of n different objects taken r at a time is
$$^nC_r = \frac{n(n-1)(n-2) \ldots (n-r+1)}{r!} = \frac{n!}{r!(n-r)!}$$

where $r! = 1 \times 2 \times 3 \times \ldots \times r$.

Note that $^nC_r = {}^nC_{n-r}$

The number of ways in which $(m + n)$ different objects can be divided into two groups of m and n things is
$$m + n\,C_n = \frac{(m+n)!}{m!n!}$$

The number of ways in which n objects can be arranged amongst themselves taken all at a time, when p are alike of one kind, q are alike of a second kind and r are alike of a third kind, the remainder being different is
$$x = \frac{n!}{p!q!r!}$$

The number of arrangements of n things taken r at a time is n^r.

Sums of numbers

The sum of the first n numbers is
$$\sum n = 1 + 2 + 3 + \ldots + n = \tfrac{1}{2}n(n+1)$$
The sum of the squares of the first n numbers is
$$\sum n^2 = 1^2 + 2^2 + 3^2 + \ldots + n^2 = \tfrac{1}{6}n(n+1)(2n+1)$$
The sum of the cubes of the first n numbers
$$\sum n^3 = 1^3 + 2^3 + 3^3 + \ldots + n^3 = \tfrac{1}{4}n^2(n+1)^2$$

Binomial theorem

This form is usually used in statistics
$$(a+b)^n = a^n + {}^nC_1 a^{n-1}b + {}^nC_2 a^{n-2}b^2 + \ldots + {}^nC_r a^{n-r}b^r + \ldots + b^n$$
$$= a^n + na^{n-1}b + \frac{n(n-1)}{2!}a^{n-2}b^2 + \ldots$$
$$+ \frac{n(n-1)(n-2)\ldots(n-r+1)}{r!}a^{n-r}b^r + \ldots + b^n$$

This form is used for approximations, etc.
$$(1 \mp x)^n = 1 + nx + \frac{n(n-1)}{2!}x^2 + \ldots$$
$$+ \frac{n(n-1)(n-2)\ldots(n-r+1)}{r!}x^r + \ldots + x^n$$

If n is a positive integer the series is finite and true for all values of x. If n is fractional or negative the series is valid only if x lies between -1 and $+1$.

Approximations from the binomial theorem

If a and b are small quantities the following relationships are approximately true:
$$(1 \pm a)^m = 1 \pm ma$$
$$(1 \pm a)^m (1 \pm b)^n = 1 \pm ma \pm nb$$

Series

The expressions in brackets following the series gives the values for x for which the series is convergent. If no information is given the series converges for all finite values of x.

Taylor's expansion

$$f(x+a) = f(x) + af'(x) + \frac{a^2}{2!}f'(x) + \frac{a^3}{3!}f''(x) + \ldots$$

Maclaurin's form

$$f(x) = f(0) + xf'(0) + \frac{x^2}{2!}f''(0) + \frac{x^3}{3!}f'''(0) + \ldots$$

Exponential

$$e = 1 + \frac{1}{1} + \frac{1}{2!} + \frac{1}{3!} + \ldots$$
$$e^x = 1 + x + \frac{x^2}{2!} + \frac{x^3}{3!} + \ldots$$
$$a^x = 1 + x \log_e a + \frac{(x \log_e a)^2}{2!} + \frac{(x \log_e a)^3}{3!} + \ldots$$

Logarithmic

$$\log_e (1+x) = x - \frac{x^2}{2} + \frac{x^3}{3} - \frac{x^4}{4} + \dots \quad \text{(for } -1 < x \leqslant 1)$$

$$\log_e (1-x) = -x - \frac{x^2}{2} - \frac{x^3}{3} - \frac{x^4}{4} - \dots \quad \text{(for } -1 \leqslant x < 1)$$

$$\log_e \left(\frac{1+x}{1-x}\right) = 2\left[x + \frac{x^3}{3} + \frac{x^5}{5} + \dots\right]$$

Trigonometric

$$\sin x = x - \frac{x^3}{3!} + \frac{x^5}{5!} - \frac{x^7}{7!} + \dots \quad x \text{ in radians}$$

$$\cos x = 1 - \frac{x^2}{2!} + \frac{x^4}{4!} - \frac{x^6}{6!} + \dots \quad x \text{ in radians}$$

$$\text{arc } \sin x = x + \frac{1}{2} \cdot \frac{x^3}{3} + \frac{1}{2} \cdot \frac{3}{4} \cdot \frac{x^5}{5} + \frac{1}{2} \cdot \frac{3}{4} \cdot \frac{5}{6} \cdot \frac{x^7}{7} + \dots$$

Hyperbolic

$$\sinh x = x + \frac{x^3}{3!} + \frac{x^5}{5!} + \dots$$

$$\cosh x = 1 + \frac{x^2}{2!} + \frac{x^4}{4!} + \dots$$

Proportion

$$\text{If } \frac{a}{b} = \frac{c}{d} \text{ then } \frac{a+b}{b} = \frac{c+d}{d}$$

$$\frac{a-b}{b} = \frac{c-d}{d}$$

$$\frac{a-b}{a+b} = \frac{c-d}{c+d}$$

Geometry

Angles

1 revolution $= 360° = 2\pi$ radians

$$60' = 1°$$

$$60'' = 1'$$

$$1° = \frac{2\pi}{360} \text{ radians}$$

$$1 \text{ radian} = \frac{360}{2\pi} = 57.3°$$

$$45° = \frac{\pi}{4} \text{ radians} \qquad 90° = \frac{\pi}{2} \text{ radians}$$

$$60° = \frac{\pi}{3} \text{ radians} \qquad 180° = \pi \text{ radians}$$

$$120° = \frac{2\pi}{3} \text{ radians} \qquad 270° = \frac{3\pi}{2} \text{ radians}$$

Types of angles

Acute angle
(less than 90°)

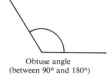

Obtuse angle
(between 90° and 180°)

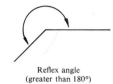

Reflex angle
(greater than 180°)

Complementary Angles are angles whose sum is 90°.

Supplementary Angles are angles whose sum is 180°.

Triangles

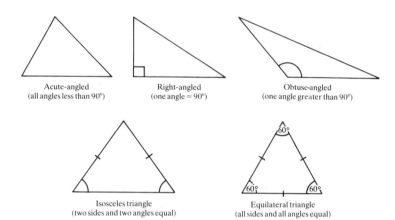

Acute-angled
(all angles less than 90°)

Right-angled
(one angle = 90°)

Obtuse-angled
(one angle greater than 90°)

Isosceles triangle
(two sides and two angles equal)

Equilateral triangle
(all sides and all angles equal)

Quadrilaterals

A *quadrilateral* is any four sided figure. The sum of its angles is 360°.

A *parallelogram* has both pairs of opposite sides parallel and it has the following properties:

(i) The sides which are opposite to each other are equal in length.

(ii) The angles which are opposite to each other are equal.

(iii) The diagonals bisect each other.

(iv) The diagonals each bisect the parallelogram.

A *rectangle* is a parallelogram with all its angles equal to 90°. It has all the properties of a parallelogram, but in addition the diagonals are equal in length.

A *rhombus* is a parallelogram with all its sides equal in length. It has all the properties of a parallelogram, but in addition it has the following properties:

(i) The diagonals bisect each other at right-angles.

(ii) The diagonal bisects the angle through which it passes.

A *square* is a rectangle with all its sides equal in length. It has all the properties of a parallelogram, rectangle and rhombus.

A *trapezium* is a quadrilateral which has one pair of sides parallel.

Polygons

Any plane figure bounded by straight lines is called a *polygon*.

A *convex* polygon has no interior angle greater than 180°.

A *re-entrant* polygon has at least one angle greater than 180°.

A *regular* polygon has all its sides and angles equal.

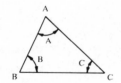

The sum of the three angles of a triangle is 180°.

$A + B + C = 180°$.

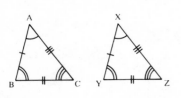

Two triangles are *congruent* if they are equal in every respect. Any of the following are sufficient to prove that two triangles are congruent:

(i) One side and two angles in one triangle equal to one side and two similarly located angles in the second triangle.

(ii) Two sides and the angle between them in one triangle equal to two sides and the angle between them in the second triangle.

(iii) Three sides in one triangle equal to three sides in the second triangle.

(iv) In right-angled triangles, the hypotenuses are equal and one side in each triangle also equal.

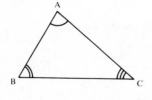

Two triangles are *similar* if they are equi-angular. If in △s ABC and XYZ, $\angle A = \angle X$, $\angle B = \angle Y$ and $\angle C = \angle Z$ then

$$\frac{AB}{XY} = \frac{AC}{XZ} = \frac{BC}{YZ}$$

Any of the following is sufficient to prove that two triangles are similar:

(i) Two angles in one triangle are equal to two angles in the second triangle.

(ii) Two sides in one triangle are proportional to two sides in the second triangle and the angle between these sides in each triangle is equal.

(iii) Three sides in one triangle are proportional to three sides in the second triangle.

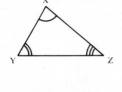

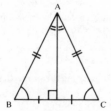

In an isosceles triangle if a line is drawn from the point where the two equal sides meet so that it bisects the third side it bisects the angle between the equal sides and it is also perpendicular to the third side.

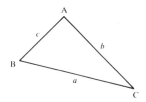

In every triangle, the greatest angle lies opposite to the longest side and the smallest angle lies opposite to the shortest side. If A is the greatest angle, a is the longest side and if C is the smallest angle c is the shortest side.

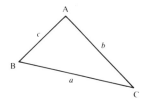

In every triangle, the sum of the lengths of two of the sides is always greater than the third side. $a+b>c$; $b+c>a$; $a+c>b$.

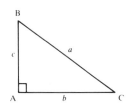

In a right-angled triangle, the square on the hypotenuse is equal to the sum of the squares on the other two sides. $a^2 = b^2 + c^2$.

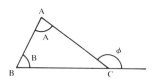

If one side of a triangle is produced, the exterior angle is equal to the sum of the two interior angles. $\phi = A + B$.

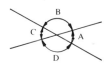

If two lines intersect the vertically opposite angles are equal. $A = C$ and $B = D$.

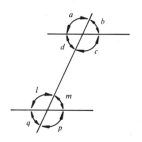

When two parallel lines are cut by a transversal:
(i) The corresponding angles are equal:
$a = l$, $b = m$, $c = p$ and $b = q$
(ii) The alternate angles are equal: $d = m$ and $c = l$
(iii) The interior angles are supplementary:
$d + l = 180°$ and $c + m = 180°$.

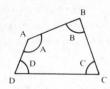

In any quadrilateral the sum of the interior angles equals 360°.
$A + B + C + D = 360°$.

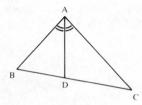

The internal bisector of the angle of a triangle divides the opposite side in the ratio of the sides containing the angle. If AD bisects the angle A then $\dfrac{AB}{AC} = \dfrac{BD}{DC}$.

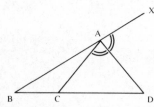

The external bisector of an angle of a triangle divides the opposite side externally in the ratio of the sides containing the angle. If AD bisects the angle CAX then $\dfrac{AB}{AC} = \dfrac{BD}{DC}$.

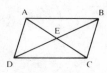

In a parallelogram the sides opposite to each other are equal in length. The angles which lie opposite to each other are equal. A diagonal divides the parallelogram into two congruent triangles. The two diagonals bisect each other. $AB = CD$; $BC = AD$; $\angle A = \angle C$; $\angle B = \angle D$; $\triangle ABC \equiv \triangle ACD$; $AE = EC$; $DE = EB$.

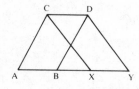

Parallelograms having equal bases and equal heights are equal in area. Area ABCD = area CXYD.

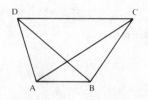

Triangles having equal bases and equal heights are equal in area. Area $\triangle ABC$ = area $\triangle ABD$.

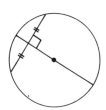

If the diameter of a circle is at right-angles to a chord then it divides the chord into two equal parts.

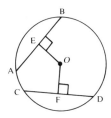

Chords which are equal in length are equi-distant from the centre of the circle. If AB = CD then $OE = OF$.

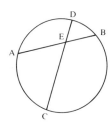

If two chords intersect inside or outside a circle the product of the segments of one chord is equal to the product of the segment of the other chord. $AE \times EB = CE \times ED$.

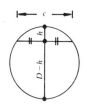

The following special case is useful

$$\frac{c^2}{4} = h(D - h) \text{ where } D = \text{diameter of the circle.}$$

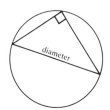

If a triangle is drawn in a semi-circle the angle opposite the diameter is a right angle.

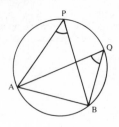

Angles in the same segment of a circle are equal. $\angle APB = \angle AQB$ because they are angles in the same segment ABQP.

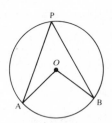

The angle which an arc of a circle subtends at the centre is twice the angle which it subtends at the circumference. $\angle AOB = 2 \times \angle APB$.

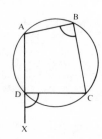

The opposite angles of a cyclic quadrilateral are supplementary (i.e. equal to 180°). $\angle A + \angle C = 180°$ and $\angle D + \angle B = 180°$. The exterior angle is equal to the interior opposite angle. That is, $\angle CDX = \angle B$, etc.

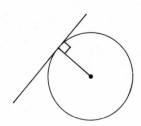

A tangent to a circle is at right-angles to a radius drawn to the point of tangency.

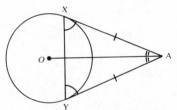

If from a point outside a circle, tangents are drawn to the circle, then the two tangents are equal in length. They also make equal angles with the chord joining the points of tangency, that is, $\angle AXY = \angle AYX$. The line drawn from the point where the tangents meet to the centre of the circle bisects the angle between the tangents, i.e. $\angle OAX = \angle OAY$.

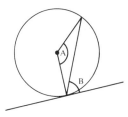

The angle between a tangent and a chord drawn from the point of tangency equals one-half of the angle at the centre subtended by the chord. Thus, $\angle B = \frac{1}{2}\angle A$.

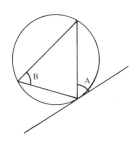

The angle between a tangent and a chord drawn from the point of tangency equals the angle at the circumference subtended by the chord. The angle at the circumference must be in the alternate segment. Thus $\angle A = \angle B$.

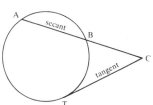

If from a point outside a circle two lines are drawn, one a secant and the other a tangent, then the square on the tangent is equal to the rectangle contained by the whole secant and that part of it which lies outside the circle. Thus, $CT^2 = AC \cdot BC$.

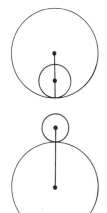

If two circles touch internally or externally, then the line which passes through their centres also passes through the point of tangency.

Geometrical constructions

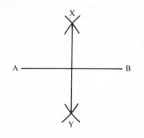

To divide a line AB into two equal parts

With A and B as centres and a radius greater than $\frac{1}{2}$AB draw circular arcs which intersect at X and Y. The line XY divides AB into two equal parts and it is also perpendicular to AB.

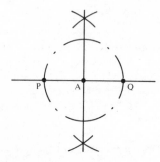

To draw a perpendicular from a given point A on a straight line

With centre A and any radius draw a circle to cut the straight line at points P and Q. With centres P and Q and a radius greater than AP (or AQ) draw circular arcs to intersect at X and Y. Join XY. This line will pass through A and is perpendicular to the given line.

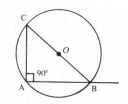

To draw a perpendicular from a point A at the end of a line

From any point O outside the line and radius OA draw a circle to cut the line at B. Draw the diameter BC and join AC. AC is perpendicular to the straight line.

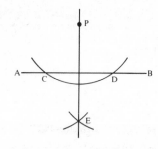

To draw the perpendicular to a line AB from a point P which does not lie on AB

With P as centre draw a circular arc to cut AB at the points C and D. With C and D as centres and a radius greater than $\frac{1}{2}$CD, draw circular arcs to intersect at E. Join PE then the line PE is the required perpendicular.

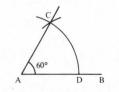

To construct an angle of 60°

Draw a line AB. With A as centre and any convenient radius draw a circular arc to cut AB at D. With D as centre and the same radius draw a second arc to cut the first arc at C. Join AC, then the angle CAD is 60°.

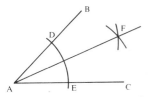

To bisect a given angle BAC

With centre A and any radius draw an arc to cut AB and AC at D and E respectively. With centres D and E and a radius greater than $\frac{1}{2}$DE, draw circular arcs to intersect at F. Join AF then AF bisects ∠BAC. (An angle of 30° is obtained by bisecting 60° and 45° is obtained by bisecting a right-angle.)

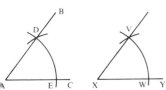

To construct an angle equal to a given angle BAC

With centre A and any radius draw an arc to cut AB at D and AC at E. Draw the line XY. With centre X and the same radius, draw an arc to cut XY at W. With centre W and a radius equal to DE draw an arc to cut the first arc at V. Join VX, then ∠VXW is equal to ∠BAC.

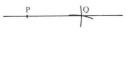

Through a point P to draw a line parallel to a given line AB

Mark off any two points X and Y on AB. With centre P and radius XY draw an arc. With centre Y and radius XP draw a second arc to cut the first arc at Q. Join PQ, then PQ is parallel to AB.

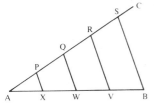

To divide a straight line AB into a number of equal parts

Suppose that AB has to be divided into four equal parts. Draw AC at any angle to AB. Set off on AC, four equal parts AP, PQ, QR and RS, of any convenient length. Join SB and draw RV, QW and PX each parallel to SB. Then AX = XW = WV = VB.

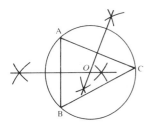

To draw the circumscribed circle of a given triangle ABC

Construct the perpendicular bisectors of the sides AB and AC so that they intersect at O. With centre O and radius AO draw a circle. This circle is the required circumscribed circle.

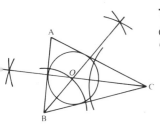

To draw the inscribed circle of a given triangle ABC

Construct the internal bisectors of the angles B and C to intersect at O. With centre O draw the inscribed circle of the triangle ABC.

17

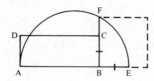

To draw a square whose area is equal to that of a given rectangle ABCD

Produce AB to E so that BC is equal to BE. Draw a circle with AE as diameter to meet BC, or BC produced, at F. Then BF is a side of the required square.

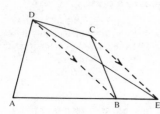

To draw a triangle whose area is equal to that of a given quadrilateral ABCD

Join BD and draw CE parallel to BD to meet AB produced at E. Then ADE is a triangle whose area is equal to the given quadrilateral ABCD.

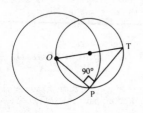

To draw a tangent to a circle at a given point P on the circumference of the circle

O is the centre of the given circle. Join OP and draw a line PT perpendicular to OP. PT is the required tangent.

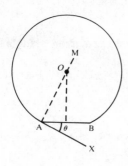

To draw the segment of a circle so that it contains the given angle θ

Draw the lines AB and AX so that ∠BAX = θ. From A draw AM perpendicular to AX. Draw the perpendicular bisector of AB to meet AM at O. With centre O and radius OA draw the circular arc which terminates at A and B. This is the required segment.

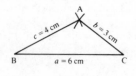

To construct a triangle given the length of the three sides

Suppose that $a = 6$ cm, $b = 3$ cm and $c = 4$ cm. Draw BC = 6 cm. With centre B and radius 4 cm draw a circular arc. With centre C and radius 3 cm draw a second arc to cut the first at A. Join AB and AC then ABC is the required triangle.

To construct a triangle given two sides and the included angle between the two sides

Suppose that $b = 5$ cm and $c = 6$ cm and $\angle A = 60°$. Draw $AB = 6$ cm and draw AX such that $\angle BAX = 60°$. Along AX mark off $AC = 5$ cm. Then ABC is the required triangle.

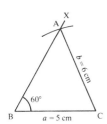

To construct a triangle (or triangles) given the lengths of two of the sides and the angle which is not included between those two sides

(a) Suppose $a = 5$ cm, $b = 6$ cm and $\angle B = 60°$. Draw $BC = 5$ cm and draw BX such that $\angle CBX = 60°$. With centre C and radius of 6 cm describe a circular arc to cut BX at A. Join CA then ABC is the required triangle.

(b) Suppose that $a = 5$ cm, $b = 4.5$ cm and $\angle B = 60°$. The construction is the same as before but the circular arc drawn with C as centre now cuts BX at two points A and A'. This means that there are two triangles which meet the given conditions. They are A'BC and ABC. This case is called the *ambiguous case*.

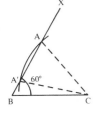

To construct a common tangent to two circles

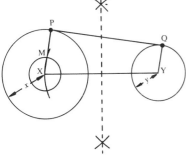

The two given circles have centres X and Y respectively and radii x and y. With centre X draw a circle whose radius is $(x - y)$. With diameter XY and centre Y draw an arc to cut the previously drawn circle at M. Join XM and produce to P at the circumference of the circle. Draw YQ parallel to XP, Q being at the circumference of the circle. Join PQ which is the required tangent.

To construct a pair of tangents from an external point to a given circle

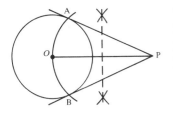

It is required to draw a pair of tangents from the point P to the circle centre O. Join OP. With OP as diameter draw a circle to cut the given circle at points A and B. Join PA and PB which are the required pair of tangents.

Mensuration

Areas of plane figures

Rectangle

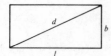

$$\text{Area} = lb = b\sqrt{d^2 - b^2} = l\sqrt{d^2 - l^2}$$
$$\text{Perimeter} = 2l + 2b$$

Square

$$\text{Area} = a^2 = \tfrac{1}{2}d^2$$
$$\text{Perimeter} = 4a$$

Parallelogram

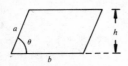

$$\text{Area} = bh = ab \sin \theta$$
$$\text{Perimeter} = 2b + 2a$$

Rhombus

$$\text{Area} = \tfrac{1}{2}cd \quad \text{(where } c \text{ and } d \text{ are the lengths of the diagonals)}$$

Trapezium

$$\text{Area} = \tfrac{1}{2}h(a + b)$$

Triangle

$$\text{Area} = \tfrac{1}{2}bh = \tfrac{1}{2}ab \sin C$$
$$= \sqrt{s(s - a)(s - b)(s - c)}$$
$$\text{where } s = \frac{a + b + c}{2}$$

adrilateral

$$\text{Area} = \frac{a(H+h)+bh+cH}{2}$$

The area of a quadrilateral may also be found by dividing the figure up into two triangles as shown by the dotted line.

olygon (regular)

$$\text{Area} = \tfrac{1}{4}nl^2 \cot \frac{180}{n}$$

(n is the number of sides of length l)

ircle

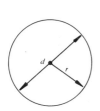

$$\text{Area} = \pi r^2 = \frac{\pi d^2}{4}$$

$$\text{Circumference} = \pi d = 2\pi r$$

egment of a circle

ector of a circle

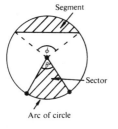

Segment

Sector

Arc of circle

$$\text{Area} = \tfrac{1}{2}r^2(\phi - \sin\phi) \ (\phi \text{ in radians})$$

$$\text{Length of chord} = 2r\sin\frac{\phi}{2}$$

$$\text{Area} = \pi r^2 \times \frac{\theta}{360} \ (\theta \text{ in degrees})$$

$$= \tfrac{1}{2}r^2\theta \ (\theta \text{ in radians}) = \tfrac{1}{2}rl$$

$$\text{Length of arc} = 2\pi r \times \frac{\theta}{360}$$

$$(\theta \text{ in degrees})$$

$$= r\theta \ (\theta \text{ in radians})$$

llipse

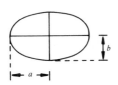

$$\text{Circumference} = 2\pi\sqrt{\tfrac{1}{2}(a^2+b^2)} \text{ approx.}$$

$$\text{Area} = \pi ab$$

arabola

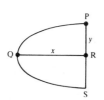

$$\text{Length of arc PQS} = 2\sqrt{y^2 + \frac{4x^2}{3}} \text{ approx.}$$

$$\text{Area of sector PQSR} = \frac{4xy}{3}$$

Inscribed and circumscribed circles

The radius of the inscribed circle of a triangle is

$$r = \frac{\sqrt{s(s-a)(s-b)(s-c)}}{s}$$

where $s = \frac{1}{2}(a+b+c)$, and a, b and c are the sides of the triangle. The radius of the circumscribed circle for a triangle is

$$r = \frac{abc}{4\sqrt{s(s-a)(s-b)(s-c)}}$$

For a regular polygon having n sides each of length l,

$$\text{radius of the inscribed circle} = \frac{l}{2}\cot\frac{180}{n}$$

$$\text{radius of the circumscribed circle} = \frac{l}{2}\operatorname{cosec}\frac{180}{n}$$

Names of polygons

Name	Number of sides	Name	Number of sides
pentagon	5	nonagon	9
hexagon	6	decagon	10
heptagon	7	undecagon	11
octagon	8	dodecagon	12

Volumes and surface areas

Cylinder

Volume $= \pi r^2 h$
Curved surface area $= 2\pi rh$
Total surface area $= 2\pi rh + 2\pi r^2$
$\qquad\qquad\qquad = 2\pi r(r+h)$

Any solid having a uniform cross-section

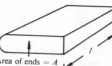

Area of ends $= A$

Volume $= Al$
Curved surface area
$\qquad = \text{perimeter of cross-section} \times \text{leng}$
Total surface area
$\qquad = \text{curved surface area} + \text{area of en}$

Cone

Volume $= \frac{1}{3}\pi r^2 h$
Curved surface area $= \pi rl$
Total surface area $= \pi rl + \pi r^2$

(h = vertical heigh
(l = slant height)

Sphere

Volume $= \frac{4}{3}\pi r^3$
Surface area $= 4\pi r^2$

rustrum of a cone

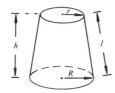

$\text{Volume} = \frac{1}{3}\pi h (R^2 + Rr + r^2)$

$\text{Curved surface area} = \pi (R + r)l$

$\text{Total surface area} = \pi (R + r)l + \pi R^2 + \pi r^2$

ramid

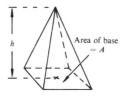

Area of base = A

$\text{Volume} = \frac{1}{3}Ah$

rism

Any solid with two faces parallel and having a constant cross-section. The end faces must be triangles, quadrilaterals or polygons.

$\text{Volume} = \text{area of cross-section} \times \text{length of prism}$

the sphere

Area of spherical triangle whose angles are A, B and C radians on a sphere whose radius is r is

$$\text{Area} = r^2(A + B + C - \pi)$$

Area of spherical polygon of n sides on a sphere of radius r is

$$\text{Area} = r^2[\theta - (n - 2)\pi]$$

where θ is the sum of its angles in radians.

Area of the curved surface is a *spherical segment* whose height is h and whose radius is r is

$$\text{Area} = 2\pi rh$$

The volume of the same spherical segment is

$$\text{Volume} = \frac{1}{3}\pi h^2(3r - h)$$
$$= \frac{1}{6}\pi h(h^2 + 3a^2)$$

where a is the radius of the base of the segment.

regular polyhedra

The table below gives the surface area and the volume of regular polyhedra in terms of the length of one edge, l.

Name	Nature of surface	Surface area	Volume
Tetrahedron	4 equilateral triangles	$1.73205l^2$	$0.11785l^2$
Cube or hexahedron	6 squares	$6l^2$	l^3
Octahedron	8 equilateral triangles	$3.46410l^2$	$0.47140l^3$
Dodecahedron	12 pentagons	$20.64578l^2$	$7.66312l^3$
Icosahedron	20 equilateral triangles	$8.66025l^2$	$2.18170l^3$

23

Trigonometry

Trigonometry

$$\sin A = \frac{\text{opposite side}}{\text{hypotenuse}} = \frac{a}{b}$$

$$\cos A = \frac{\text{adjacent side}}{\text{hypotenuse}} = \frac{c}{b}$$

$$\tan A = \frac{\text{opposite side}}{\text{adjacent side}} = \frac{a}{c}$$

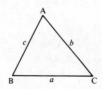

$$\operatorname{cosec} A = \frac{1}{\sin A} = \frac{\text{hypotenuse}}{\text{opposite side}} = \frac{b}{a}$$

$$\sec A = \frac{1}{\cos A} = \frac{\text{hypotenuse}}{\text{adjacent side}} = \frac{b}{c}$$

$$\cot A = \frac{1}{\tan A} = \frac{\text{adjacent side}}{\text{opposite side}} = \frac{c}{a}$$

$$\sin 60° = \frac{\sqrt{3}}{2} \quad \sin 30° = \frac{1}{2} \quad \sin 45° = \frac{\sqrt{2}}{2}$$

$$\cos 60° = \frac{1}{2} \quad \cos 30° = \frac{\sqrt{3}}{2} \quad \cos 45° = \frac{\sqrt{2}}{2}$$

$$\tan 60° = \sqrt{3} \quad \tan 30° = \frac{\sqrt{3}}{3} \quad \tan 45° = 1$$

$$\cos A = \sin(90° - A)$$
$$\sin A = \cos(90° - A)$$

Trigonometrical identities

$$\sin^2 A + \cos^2 A = 1 \qquad \sec^2 A = 1 + \tan^2 A$$

$$\operatorname{cosec}^2 A = 1 + \cot^2 A \qquad \tan A = \frac{\sin A}{\cos A}$$

The general angle

Quadrant	Angle	$\sin A =$	$\cos A =$	$\tan A =$
first	0 to 90°	$\sin A$	$\cos A$	$\tan A$
second	90° to 180°	$\sin(180° - A)$	$-\cos(180° - A)$	$-\tan(180° - A)$
third	180° to 270°	$-\sin(A - 180°)$	$-\cos(A - 180°)$	$\tan(A - 180°)$
fourth	270° to 360°	$-\sin(360° - A)$	$\cos(360° - A)$	$-\tan(360° - A)$

For any triangle

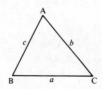

Sine rule

$$\frac{a}{\sin A} = \frac{b}{\sin B} = \frac{c}{\sin C}$$

Cosine rule

$$a^2 = b^2 + c^2 - 2bc \cos A$$
$$b^2 = a^2 + c^2 - 2ac \cos B$$
$$c^2 = a^2 + b^2 - 2ab \cos C$$

Tangent rule

$$\tan \frac{B-C}{2} = \frac{b-c}{b+c} \cot \frac{A}{2}$$

Cotangent rule

$$(m+n) \cot \theta = m \cot \alpha - n \cot \beta$$
$$(m+n) \cot \theta = n \cot A - m \cot B$$

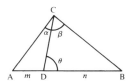

Half-angle formulae

$$\sin \frac{A}{2} = \sqrt{\frac{(s-b)(s-c)}{bc}}$$

$$\cos \frac{A}{2} = \sqrt{\frac{s(s-a)}{bc}}$$

$$\tan \frac{A}{2} = \sqrt{\frac{(s-b)(s-c)}{s(s-a)}}$$

where $s = \frac{1}{2}(a+b+c)$

Factor formulae

$$\cos P + \cos Q = 2 \cos \frac{P+Q}{2} \cos \frac{P-Q}{2}$$

$$\cos P - \cos Q = -2 \sin \frac{P+Q}{2} \sin \frac{P-Q}{2}$$

$$\sin P + \sin Q = 2 \sin \frac{P+Q}{2} \cos \frac{P-Q}{2}$$

$$\sin P - \sin Q = 2 \cos \frac{P+Q}{2} \sin \frac{P-Q}{2}$$

Multiple angles

$$\sin (A \pm B) = \sin A \cos B \pm \cos A \sin B$$
$$\cos (A \pm B) = \cos A \cos B \mp \sin A \sin B$$

$$\tan (A \pm B) = \frac{\tan A \pm \tan B}{1 \mp \tan A \tan B}$$

$$R \sin (\omega t \pm \alpha) = a \sin \omega t \pm b \cos \omega t$$

where $R = \sqrt{a^2 + b^2}$ and $\tan \alpha = \frac{b}{a}$

Spherical triangle

If A, B and C are the three angles and a, b and c the corresponding opposite sides:

$$\frac{\sin A}{\sin a} = \frac{\sin B}{\sin b} = \frac{\sin C}{\sin c}$$

$$\cos a = \cos b \cos c + \sin b \sin c \cos A$$

$$\cos A = -\cos B \cos C + \sin B \sin C \cos a$$

$$\sin \tfrac{1}{2}A = \sqrt{\frac{\sin (s-b) \sin (s-c)}{\sin b \sin c}} = \sqrt{\frac{\cos S \cos (S-A)}{-\sin B \sin C}}$$

where $s = \tfrac{1}{2}(a+b+c)$ and $S = \tfrac{1}{2}(A+B+C)$

$$\cos \tfrac{1}{2}A = \sqrt{\frac{\sin s \sin (s-a)}{\sin b \sin c}} = \sqrt{\frac{\cos (S-B) \cos (S-C)}{\sin B \sin C}}$$

$$\tan \tfrac{1}{2}A = \frac{r}{\sin (s-a)} = R \cos (S-A)$$

where $r = \sqrt{\dfrac{\sin (s-a) \sin (s-b) \sin (s-c)}{\sin s}}$

and $R = \sqrt{\dfrac{-\cos S}{\cos (S-A) \cos (S-B) \cos (S-C)}}$

Calculus

Function	Derivative	Function	Derivative
ax^n	anx^{n-1}	$\cot x$	$-\operatorname{cosec}^2 x$
$\sin ax$	$a \cos ax$	$\sec x$	$\tan x \sec x$
$\cos ax$	$-a \sin ax$	$\operatorname{cosec} x$	$-\cot x \operatorname{cosec} x$
$\tan ax$	$a \sec^2 ax$	$\arcsin x$	$(1-x^2)^{-1/2}$
$\log_e ax$	$\dfrac{1}{x}$	$\arccos x$	$-(1-x^2)^{-1/2}$
e^{ax}	ae^{ax}	$\arctan x$	$(1+x^2)^{-1}$

Product rule: If $y = uv$, $\dfrac{dy}{dx} = v\dfrac{du}{dx} + u\dfrac{dv}{dx}$

Quotient rule: If $y = \dfrac{u}{v}$, $\dfrac{dy}{dx} = \dfrac{v\dfrac{du}{dx} - u\dfrac{dv}{dx}}{v^2}$

Function of a function: $\dfrac{dy}{dx} = \dfrac{dy}{du} \cdot \dfrac{du}{dx}$

Conditions for maximum and minimum:

y has a maximum value if $\dfrac{dy}{dx} = 0$ and $\dfrac{d^2y}{dx^2}$ is negative

y has a minimum value if $\dfrac{dy}{dx} = 0$ and $\dfrac{d^2y}{dx^2}$ is positive

Point of inflexion: $\dfrac{d^2y}{dx^2}$ is zero and changes sign

**elocity and
cceleration**

If $s = f(t)$

$$v = \frac{ds}{dt}$$

$$a = \frac{d^2 s}{dt^2} = \frac{dv}{dt}$$

s = distance travelled
v = velocity
a = acceleration

Function	Integral
x^n	$\dfrac{x^{n+1}}{n+1}$
$\dfrac{1}{x}$	$\log_e x$
e^x	e^x
$\log x$	$x \log x - x$
$\dfrac{1}{a^2 + x^2}$	$\dfrac{1}{a} \arctan \dfrac{x}{a}$
$\dfrac{1}{a^2 - x^2}$	$\dfrac{1}{2a} \log \dfrac{a+x}{a-x} = \dfrac{1}{a} \operatorname{arctanh} \dfrac{x}{a}$
$\dfrac{1}{\sqrt{a^2 - x^2}}$	$\arcsin \dfrac{x}{a}$
$\sin x$	$-\cos x$
$\cos x$	$\sin x$
$\tan x$	$-\log \cos x$
$\cot x$	$\log \sin x$
$\sec x$	$\log \tan \left(\dfrac{\pi}{4} + \dfrac{x}{2} \right)$
$\operatorname{cosec} x$	$\log \tan \tfrac{1}{2} x$

**ntegration by
arts**

$$\int u \frac{dv}{dx} \, dx = uv - \int v \frac{du}{dx} \, dx$$

rea

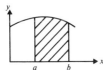

Area $A = \displaystyle\int_a^b y \, dx$

Volume generated by rotating the area A completely about the x-axis is $\pi \displaystyle\int_a^b y^2 \, dx$.

**Centroid of a
lane area**

$$\bar{x} = \frac{\displaystyle\int_a^b xy \, dx}{\text{area}} \qquad \bar{y} = \frac{\tfrac{1}{2}\displaystyle\int_a^b y^2 \, dx}{\text{area}}$$

**Second moment
f area**

$$I \text{ (about } y\text{-axis)} = \int_a^b x^2 y \, dx$$

$I = Ak^2$ where A is the area and k is the radius of gyration

Theorem of perpendicular axes

If OX and OY are two axes in the plane of a lamina and OZ is mutuall perpendicular to OX and OY then

$$I_{OZ} = I_{OX} + I_{OY}$$

Theorem of parallel axes

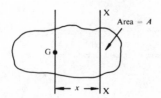

$$I_{XX} = I_G + Ax^2$$

Pappus' Theorem

Surface area of revolution $= 2\pi\bar{y}L$
Volume of revolution $\quad = 2\pi\bar{y}A$

Irregular plane areas

Mid-ordinate rule: Area $= b(h_1 + h_2 + h_3 + \ldots h_n)$ see Fig. 1

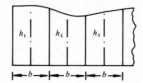 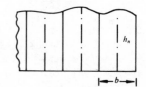

Trapezium (Trapezoidal) rule: Area $= \dfrac{b}{2}\left[(h_0 + h_n) + 2(h_1 + h_2 + \ldots h_{n-1})\right]$ se Fig. 2

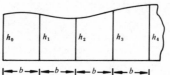 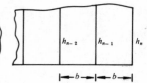

Simpson's rule: Area $= \dfrac{b}{3}\left[(h_0 + h_n) + 4(h_1 + h_3 + \ldots h_{n-1}) + 2(h_2 + h_4 + \ldots h_{n-2})\right]$

(n must be even)

Velocity and acceleration

If $a = f(t)$

a = acceleration
v = velocity
s = distance moved

$$v = \int a \, dt$$

$$s = \int v \, dt$$

Differential Equations

(1) $\dfrac{dy}{dx} = f(x)$, or $P\dfrac{dy}{dx} + Q = 0$, P and Q being functions of x only.

$P\dfrac{dy}{dx} + Q = 0$ can be written $\dfrac{dy}{dx} = -\dfrac{Q}{P} = f(x)$.

SOLUTION: $y = \displaystyle\int f(x)\,dx + c$

(2) $\dfrac{dy}{dx} = F(y)$, or $P\dfrac{dy}{dx} + Q = 0$, P and Q being functions of y only.

$P\dfrac{dy}{dx} + Q = 0$ can be written $\dfrac{dy}{dx} = -\dfrac{Q}{P} = F(y)$.

SOLUTION: $x = \displaystyle\int \dfrac{1}{F(y)}\,dy + c$

(3) $\dfrac{d^n y}{dx^n} = f(x)$, then $\dfrac{d^{n-1}y}{dx^{n-1}} = \displaystyle\int f(x)\,dx + c_1$

$\dfrac{d^{n-2}y}{dx^{n-2}} = \displaystyle\int \left(\int f(x)\,dx + c_1 \right) dx + c_2$ etc.

SOLUTION: Obtained by continuous integration until y is obtained, each integration introduces an arbitrary constant.

(4) $P\dfrac{dy}{dx} + Q = 0$, where P is a function of y and Q is a function of x (or the equation can be written in this form).

SOLUTION: $\displaystyle\int P\,dy = -\int Q\,dx + c$

(5) $P\dfrac{dy}{dx} + Q = 0$, P and Q being functions of x and y; if this can be written

$P\,dy + Q\,dx = 0$, where the left hand side of the equation is an exact differential

then $\dfrac{\partial Q}{\partial y} = \dfrac{\partial P}{\partial x}$.

SOLUTION: If the equation is exact the integral can be written down directly.

(6) $\dfrac{dy}{dx} + Py = Q$, where P and Q are functions of x only. This can be made exact by multiplying by the integrating factor $e^{\int P\,dx}$.

SOLUTION: Multiply both sides of the equation by the integrating factor $e^{\int P\,dx}$. The left hand side will then be an exact differential and the right hand side can be integrated in the normal way.

(7) $\dfrac{d^2 y}{dx^2} = f(y)$, this can be written $p\dfrac{dp}{dx} = f(y)$ by putting $p = \dfrac{dy}{dx}$.

SOLUTION: $\displaystyle\int p\,dp = \int f(y)\,dy + c$

(8) $\dfrac{d^2y}{dx^2} + a\dfrac{dy}{dx} + by = 0,$ the solution depends on the form of the roots of the

equation $k^2 + ak + b = 0,$ called the auxiliary equation.

SOLUTION:

    (1)   Roots of auxiliary equation real and different, say α and β

$$y = A\,e^{\alpha x} + B\,e^{\beta x}$$

    (2)   Roots of the auxiliary equation equal, say α

$$y = (Ax + B)\,e^{\alpha x}$$

    (3)   Roots of the auxiliary equation complex, say $p + jq,$ $p - jq$

$$y = e^{px}(C\cos qx + D\sin qx)$$

    or $y = R\,e^{px}\sin(qx + \theta),$ where R and θ are the arbitrary constants.

(9) $\dfrac{d^2y}{dx^2} + a\dfrac{dy}{dx} + by = f(x).$

SOLUTION: The general solution of this type of equation is made up of two parts, the Complementary Function, as above (8), and the Particular Integral, which depends on the $f(x)$ and can be found by using the D operator method.

NOTE: An alternative method of solution of this type of equation when the boundary conditions are given is to use Laplace Transforms.

Definition: The Laplace Transform of a function of t, say $F(t)$,   is

$\mathscr{L}\{F(t)\},$ where $\mathscr{L}\{F(t)\} = \displaystyle\int_{o}^{\infty} e^{-st}F(t)\,dt.$

Function	Laplace Transform	Function	Laplace Transform
1	$\dfrac{1}{s}$	t^n	$\dfrac{n!}{s^{n+1}}$
e^{at}	$\dfrac{1}{s-a}$	$\sinh at$	$\dfrac{a}{s^2-a^2}$
$\sin at$	$\dfrac{a}{s^2+a^2}$	$\cosh at$	$\dfrac{s}{s^2-a^2}$
$\cos at$	$\dfrac{s}{s^2+a^2}$	$e^{-at}t^n$	$\dfrac{n!}{(s+a)^{n+1}}$
t	$\dfrac{1}{s^2}$	$e^{-at}\cos at$	$\dfrac{s+b}{(s+b)^2+a^2}$

Special cases

(10) $\dfrac{d^2y}{dx^2} = k_1 + k_2 x + k_3 x^2 + k_4 x^3 + \ldots + k_{n-1}x^{n-2}$

SOLUTION: $y = k + k_0 x + k\dfrac{x^2}{1\times 2} + k\dfrac{x^3}{2\times 3} + \ldots + k_{n-1}\dfrac{x^n}{n(n-1)}$

This equation is used in the deflection of beams.

(11) $\dfrac{d^2y}{dt^2}+\omega^2 y=0,$ Simple Harmonic Motion

y is the distance moved in time t and ω is the frequency in Hertz.

SOLUTION: $y=y_0\sin(\omega t+\alpha)$ y_0 is the amplitude and α is the phase angle.

(12) $\dfrac{d^2y}{dt^2}+2k\dfrac{dy}{dt}+\omega^2 y=0$ Damped Vibrations. $2k$ is the damping coefficient.

SOLUTION: $k<\omega,\ y=y_0\,e^{-kt}\sin(\sqrt{\omega^2-k^2}\ldots t+\alpha)$

$k=\omega,\ y=(A+Bt)\,e^{-kt}$

$k>\omega,\ y=e^{-kt}(A\ e^{\sqrt{k^2-\omega^2}}\ldots t+B\ e^{\sqrt{-k^2-\omega^2}}\ldots t)$

Centroids of some plane figures

Rectangle

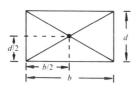

Centroid lies at the intersection of the diagonals.

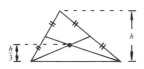

Centroid lies at the intersection of the bisectors of the sides.

Semi-circle

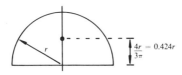

Quadrant of a circle

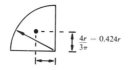

Sector of a circle

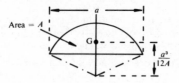

Segment of a circle

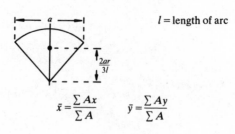

l = length of arc

$$\bar{x} = \frac{\sum Ax}{\sum A} \qquad \bar{y} = \frac{\sum Ay}{\sum A}$$

Hyperbolic functions

$$\sinh x = \tfrac{1}{2}(e^x - e^{-x})$$

$$\cosh x = \tfrac{1}{2}(e^x + e^{-x})$$

$$\tanh x = \frac{\sinh x}{\cosh x} = \frac{e^{2x} - 1}{e^{2x} + 1}$$

$$\operatorname{cosech} x = \frac{1}{\sinh x} = \frac{2}{e^x - e^{-x}}$$

$$\operatorname{sech} x = \frac{1}{\cosh x} = \frac{2}{e^x + e^{-x}}$$

$$\coth x = \frac{1}{\tanh x} = \frac{e^{2x} + 1}{e^{2x} - 1}$$

$$\cosh^2 x - \sinh^2 x = 1$$

$$\operatorname{sech}^2 x = 1 - \tanh^2 x$$

$$\operatorname{cosech}^2 x = \coth^2 x - 1$$

$$\sinh (x \pm y) = \sinh x \cosh y \pm \cosh x \sinh y$$

$$\cosh (x \pm y) = \cosh x \cosh y \pm \sinh x \sinh y$$

$$\tanh (x \pm y) = \frac{\tanh x \pm \tanh y}{1 \pm \tanh x \tanh y}$$

$$\operatorname{arc\,sinh} x = \log_e (x + \sqrt{x^2 + 1})$$

$$\operatorname{arc\,cosh} x = \pm\log_e (x + \sqrt{x^2 - 1})$$

$$\operatorname{arc\,tanh} x = \tfrac{1}{2} \log_e \frac{1 + x}{1 - x}$$

Complex numbers

If x and y are real numbers and $i = \sqrt{-1}$ then the complex number $z = x + iy$ consists of the real part x and the imaginary part iy.

$\bar{z} = x - iy$ is the conjugate of the complex number $z = x + iy$.

If $x + iy = a + ib$ then $x = a$ and $y = b$

$$(a + ib) + (c + id) = (a + c) + i(b + d)$$
$$(a + ib) - (c + id) = (a - c) + i(b - d)$$
$$(a + ib)(c + id) = (ac - bd) + i(ad + bc)$$
$$\frac{a + ib}{c + id} = \frac{ac + bd}{c^2 + d^2} + i\frac{bc - ad}{c^2 + d^2}$$

Every complex number may be written in polar form. Thus

$$x + iy = r(\cos\theta + i\sin\theta) = r\angle\theta$$

r is called the modulus of z and this may be written $r = |z|$

$$r = \sqrt{x^2 + y^2}$$

θ is called the argument and this may be written $\theta = \arg z$

$$\tan\theta = \frac{y}{x}$$

If $z_1 = r_1(\cos\theta_1 + i\sin\theta_1)$ and $z_2 = r_2(\cos\theta_2 + i\sin\theta_2)$

$$z_1 z_2 = r_1 r_2[\cos(\theta_1 + \theta_2) + i\sin(\theta_1 + \theta_2)] = r_1 r_2\angle(\theta_1 + \theta_2)$$

$$\frac{z_1}{z_2} = \frac{r_1[\cos(\theta_1 - \theta_2) + i\sin(\theta_1 + \theta_2)]}{r_2} = \frac{r_1}{r_2}\angle(\theta_1 - \theta_2)$$

Exponential form

$$z = r\,e^{i\theta}$$

De Moivre's theorem

$$(\cos\theta + i\sin\theta)^n = \cos n\theta + i\sin n\theta$$

where n is any real number

$$\sin\theta = \frac{e^{i\theta} - e^{-i\theta}}{2i}$$

$$\cos\theta = \frac{e^{i\theta} + e^{-i\theta}}{2}$$

$$\sin i\theta = i\sinh\theta \qquad \sinh i\theta = i\sin\theta$$
$$\cos i\theta = \cosh\theta \qquad \cosh i\theta = \cos\theta$$

Co-ordinate geometry

The straight-line

General equation	$ax + by + c = 0$	
Gradient equation	$y = mx + c$	
Intercept equation	$\dfrac{x}{A} + \dfrac{y}{B} = 1$	
Perpendicular equation	$x\cos\alpha + y\sin\alpha = p$	

m = gradient
c = intercept on the y-axis

A = intercept on the x-axis
B = intercept on the y-axis

p = length of perpendicular from the origin to the line

α = angle that the perpendicular makes with the x-axis

The distance between the points $P(x_1, y_1)$ and $Q(x_2, y_2)$ is given by

$$PQ = \sqrt{(x_1 - x_2)^2 + (y_1 - y_2)^2}$$

The equation of the line joining the points (x_1, y_1) and (x_2, y_2) is

$$\frac{y - y_1}{y_1 - y_2} = \frac{x - x_1}{x_1 - x_2}$$

The straight-line having a gradient m which passes through the point (x_1, y_1) ha the equation

$$y - y_1 = m(x - x_1)$$

If two straight lines have gradients m_1 and m_2 the angle θ between the lines is give by

$$\tan \theta = \frac{m_1 - m_2}{1 + m_1 m_2}$$

If $m_1 = m_2$ the lines are parallel.

If $m_1 m_2 = -1$ the lines are perpendicular.

The area of a triangle ABC with vertices $A(x_1, y_1)$, $B(x_2, y_2)$ and $C(x_3, y_3)$ is give by

$$\text{Area } \triangle ABC = \tfrac{1}{2}|x_1(y_2 - y_3) + x_2(y_3 - y_1) + x_3(y_1 - y_2)|$$

The circle

General equation $x^2 + y^2 + 2gx + 2fy + c = 0$.

The centre has co-ordinates $(-g, -f)$.

The radius is $r = \sqrt{g^2 + f^2 - c}$.

The equation of the tangent at (x_1, y_1) to the circle is

$$xx_1 + yy_1 + g(x + x_1) + f(y + y_1) + c = 0$$

The length of the tangent from (x_1, y_1) to the circle is

$$t^2 = x_1^2 + y_1^2 + 2gx_1 + 2fy_1 + c$$

Conic sections

The parabola

Eccentricity $= e = \dfrac{SP}{PD} = 1$

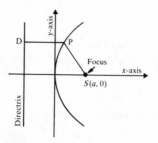

With focus $S(a, 0)$ the equation of a parabola is $y^2 = 4ax$

The parametric form of the equation is $x = at^2$, $y = 2at$

The equation of the tangent at (x_1, y_1) is $yy_1 = 2a(x + x_1)$

The ellipse

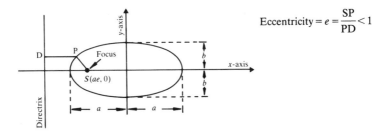

$$\text{Eccentricity} = e = \frac{SP}{PD} < 1$$

The equation of an ellipse is $\frac{x^2}{a^2} + \frac{y^2}{b^2} = 1$ where $b^2 = a^2(1 - e^2)$.

The equation of the tangent at (x_1, y_1) is $\frac{xx_1}{a^2} + \frac{yy_1}{b^2} = 1$.

The parametric form of the equation of an ellipse is $x = a \cos \theta$, $y = b \sin \theta$, where θ is the eccentric angle.

Hyperbola

$$\text{Eccentricity} = e = \frac{SP}{PD} > 1.$$

The equation of a hyperbola is $\frac{x^2}{a^2} - \frac{y^2}{b^2} = 1$ where $b^2 = a^2(e^2 - 1)$.

The parametric form of the equation is $x = a \sec \theta$, $y = b \tan \theta$ where θ is the eccentric angle.

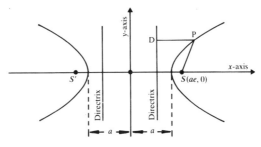

The equation of the tangent at (x_1, y_1) is $\frac{xx_1}{a^2} - \frac{yy_1}{b^2} = 1$.

Graphs

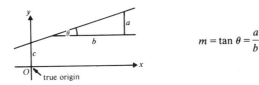

$$m = \tan \theta = \frac{a}{b}$$

The equation of a straight line can be written in the form $y = mx + c$ where m is the gradient of the line and c is the intercept on the y-axis.

Non-linear relationships can sometimes be converted into linear relationships. The most common of these are given in the table below:

35

Equation	Plot	Gradient	Intercept
$y = ax^n + b$	$y \ \nu \ x^n$	a	b
$y = \dfrac{a}{x^n} + b$	$y \ \nu \ \dfrac{1}{x^n}$	a	b
$y = a\sqrt{x} + b$	$y \ \nu \ \sqrt[n]{x}$	a	b
$y = ax^n + bx^{n-1}$	$\dfrac{y}{x^{n-1}} \ \nu \ x$	b	b
$y = ax^n$	$\log y \ \nu \ \log x$	n	$\log a$
$y = ab^x$	$\log y \ \nu \ x$	$\log b$	$\log a$
$y = a\,e^{bx}$	$\log y \ \nu \ x$	$b \log e$	$\log a$

Variation

If $y \propto x$ then $y = kx$. This is direct variation.

If $y \propto \dfrac{1}{x}$ then $y = \dfrac{k}{x}$. This is inverse variation.

If p varies directly as t and inversely as v then $p = \dfrac{kt}{v}$. This is joint variation.

The function $(ax + bx^2)$ is the sum of two parts, ax which varies directly as x and bx^2 which varies directly as x^2.

Sets

Two sets are equal if every member of one set is a member of the other set.

$A \subset B$ means A is a subset of B

$a \in A$ means a is an element of A

$a \notin A$ means a is not an element of A

$A \cap B$ means the intersection of set A and set B

$A \cup B$ means the union of set A and set B

$\mathscr{E}$ is the universal set from which subsets are formed

$A - B$ means the complement of a set A

Intersection

If

then

$A = \{a, b, c, d, e\}$ and $B = \{b, e, f, g, h, k, j\}$

$A \cap B = \{b, e\}$

Union

If

then

$A = \{a, b, c, d, e\}$ and $B = \{e, f, g\}$

$A \cup B = \{a, b, c, d, e, f, g\}$

Complement

If

then

$\mathscr{E} = \{a, b, c, d, e, f\}$ and $A = \{a, b, f\}$

$A' = \{c, d, e\}$

Difference of two sets

$A - B = A \cap B'$ $(A \cap B)' = A' \cup B'$

The number of elements in a set	$n(A \cup B) = n(A) + n(B) - n(A \cap B)$ $n(A) =$ number of elements in A etc.

$$n(A \cup B \cup C) = n(A) + n(B) + n(C) - n(A \cap B) - n(A \cap C)$$
$$- n(B \cap C) + n(A \cap B \cap C)$$

Laws for intersections and unions

(1) $A \cup B = B \cup A$

(2) $A \cup (B \cup C) = (A \cup B) \cup C$

(3) $A \cap (B \cup C) - (A \cap B) \cup (A \cap C)$

Boolean algebra

Series

X *and* Y is represented by $X \cdot Y$ or $X \cap Y$

Commutative Law	$X \cdot Y = Y \cdot X$	$X \cap Y = Y \cap X$
Idempotent Law	$X \cdot X \cdot X \ldots = X$	$X \cap X \cap X \ldots = X$
(In particular	$X \cdot 0 = 0 \cdot X = 0$	$X \cap 0 = 0 \cap X = 0$
	$X \cdot 1 = 1 \cdot X = X$	$X \cap 1 = 1 \cap X = X)$
Associative Law	$(X \cdot Y) \cdot Z = X \cdot (Y \cdot Z)$	$(X \cap Y) \cap Z = X \cap (Y \cap Z)$

Parallel

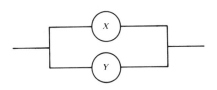

X *or* Y is represented by $X + Y$ or $X \cup Y$

Commutative Law	$X + Y = Y + X$	$X \cup Y = Y \cup X$
Idempotent Law	$X + X + X + \ldots = X$	$X \cup X \cup X \ldots = X$
(In particular	$0 + X = X + 0 = X$	$0 \cup X = X \cup 0 = X$
	$1 + X = X + 1 = 1$	$1 \cup X = X \cup 1 = 1)$
Associative Law	$(X + Y) + Z = X + (Y + Z)$	$(X \cup Y) \cup Z = X \cup (Y \cup Z)$

The complement of X is X'

X	0	1
X'	1	0

$$X \cdot X' = X' \cdot X = 0 \qquad X \cap X' = X' \cap X = 0$$
$$X + X' = X' + X = 1 \qquad X \cup X' = X' \cup X = 1$$

Distributive Laws

(1) $X \cdot (Y + Z) = (X \cdot Y) + (X \cdot Z)$ $X \cap (Y \cup Z) = (X \cap Y) \cup (X \cap Z)$

(2) $X + (Y \cdot Z) = (X + Y) \cdot (X + Z)$ $X \cup (Y \cap Z) = (X \cup Y) \cap (X \cup Z)$

Determinants

If a, b, c and d are any four numbers then

$$\Delta = \begin{vmatrix} a & b \\ c & d \end{vmatrix} = ad - bc$$

Δ is called the determinant of order 2.

The determinant of order 3

$$\Delta = \begin{vmatrix} a_{11} & a_{12} & a_{13} \\ a_{21} & a_{22} & a_{23} \\ a_{31} & a_{32} & a_{33} \end{vmatrix} = a_{11}\begin{vmatrix} a_{22} & a_{23} \\ a_{32} & a_{33} \end{vmatrix} - a_{12}\begin{vmatrix} a_{21} & a_{23} \\ a_{31} & a_{33} \end{vmatrix}$$

$$+ a_{13}\begin{vmatrix} a_{21} & a_{22} \\ a_{31} & a_{32} \end{vmatrix}$$

$$= a_{11}(a_{22}a_{33} - a_{23}a_{32}) - a_{12}(a_{21}a_{33} - a_{23}a_{31})$$

$$+ a_{13}(a_{21}a_{32} - a_{22}a_{31})$$

Minor

The minor of an element a_{ij} is the determinant of order $(n-1)$ formed from Δ by omitting the row and column containing a_{ij}.

Cofactor

The cofactor of an element a_{ij} is its minor with the sign $+$ or $-$ according to the formula

$$A_{ij} = (-1)^{i+j} \times (\text{minor of } a_{ij})$$

Value of a determinant

The value of a determinant of order n is

$$\Delta = a_{11}A_{11} + a_{12}A_{12} + a_{13}A_{13} + \ldots + a_{1n}A_{1n}$$

Using determinants to solve equations

To solve the equations

$$a_{11}x_1 + a_{12}x_2 + \ldots + a_{1n}x_n = b_1$$
$$a_{21}x_1 + a_{22}x_2 + \ldots + a_{2n}x_n = b_2$$
$$\vdots \qquad \vdots \qquad \cdots \qquad \vdots \qquad \vdots$$
$$a_{n1}x_1 + a_{n2}x_2 + \ldots + a_{nn}x_n = b_n$$

$$\text{Let } \Delta = \begin{vmatrix} a_{11} & a_{12} & \cdots & a_{1n} \\ a_{21} & a_{22} & \cdots & a_{2n} \\ \vdots & \vdots & \cdots & \vdots \\ a_{n1} & a_{n2} & \cdots & a_{nn} \end{vmatrix}$$

$$\text{then } \Delta x_j = \div \begin{vmatrix} a_{11} & a_{12} & \cdots & a_{1j-1} & b_1 & a_{1j+1} & \cdots & a_{1n} \\ a_{21} & a_{22} & \cdots & a_{2j-1} & b_2 & a_{2j+1} & \cdots & a_{2n} \\ \vdots & \vdots & \cdots & \vdots & \vdots & \vdots & \cdots & \vdots \\ a_{n1} & a_{n2} & \cdots & a_{nj-1} & b_n & a_{nj+1} & \cdots & a_{nn} \end{vmatrix}$$

This method of solving linear equations is recommended only for small values of n.

Matrices

A matrix which has an array of $m \times n$ numbers arranged in m rows and n columns is called an $m \times n$ matrix. It is denoted by

$$\begin{pmatrix} a_{11} & a_{12} & \cdots & a_{1n} \\ a_{21} & a_{22} & \cdots & a_{2n} \\ \vdots & \vdots & \vdots & \vdots \\ a_{m1} & a_{m2} & \cdots & a_{mn} \end{pmatrix}$$

Row matrix

This is a matrix having only 1 row. Thus $(a_{11} \ a_{12} \cdots a_{1n})$ is a row matrix of order $1 \times n$.

Column matrix

This is a matrix having only 1 column. Thus:

$$\begin{pmatrix} a_{11} \\ a_{21} \\ \vdots \\ a_{m1} \end{pmatrix}$$ is a column matrix of order $m \times 1$.

Null matrix

This is a matrix with all its elements zero.

$\begin{pmatrix} 0 & 0 \\ 0 & 0 \end{pmatrix}$ is a null matrix of order 2×2.

Square matrix

This is a matrix having the same number of rows and columns.

$\begin{pmatrix} a_{11} & a_{12} & a_{13} \\ a_{21} & a_{22} & a_{23} \\ a_{31} & a_{32} & a_{33} \end{pmatrix}$ is a square matrix of order 3×3.

Diagonal matrix

This is a square matrix in which all the elements are zero except those in the leading diagonal.

$\begin{pmatrix} a_{11} & 0 & 0 \\ 0 & a_{22} & 0 \\ 0 & 0 & a_{33} \end{pmatrix}$ is a diagonal matrix of order 3×3.

Unit matrix

This is a diagonal matrix with the elements in the leading diagonal all equal to 1. All other elements are 0. The unit matrix is denoted by I.

$$I = \begin{pmatrix} 1 & 0 & 0 \\ 0 & 1 & 0 \\ 0 & 0 & 1 \end{pmatrix}$$

Addition of matrices

Two matrices may be added provided that they are of the same order. This is done by adding the corresponding elements in each matrix.

$$\begin{pmatrix} a_{11} & a_{12} & a_{13} \\ a_{21} & a_{22} & a_{23} \end{pmatrix} + \begin{pmatrix} b_{11} & b_{12} & b_{13} \\ b_{21} & b_{22} & b_{23} \end{pmatrix} = \begin{pmatrix} a_{11}+b_{11} & a_{12}+b_{12} & a_{13}+b_{13} \\ a_{21}+b_{21} & a_{22}+b_{22} & a_{23}+b_{23} \end{pmatrix}$$

Subtraction of matrices

Subtraction is done in a similar way to addition except that the corresponding elements are subtracted.

$$\begin{pmatrix} a_{11} & a_{12} \\ a_{21} & a_{22} \end{pmatrix} - \begin{pmatrix} b_{11} & b_{12} \\ b_{21} & b_{22} \end{pmatrix} = \begin{pmatrix} a_{11}-b_{11} & a_{12}-b_{12} \\ a_{21}-b_{21} & a_{22}-b_{22} \end{pmatrix}$$

Scalar multiplication

A matrix may be multiplied by a number as follows:

$$b\begin{pmatrix} a_{11} & a_{12} \\ a_{21} & a_{22} \end{pmatrix} = \begin{pmatrix} ba_{11} & ba_{12} \\ ba_{21} & ba_{22} \end{pmatrix}$$

General matrix multiplication

Two matrices can be multiplied together provided the number of columns in the first matrix is equal to the number of rows in the second matrix.

$$\begin{pmatrix} a_{11} & a_{12} & a_{13} \\ a_{21} & a_{22} & a_{23} \end{pmatrix} \begin{pmatrix} b_{11} & b_{12} \\ b_{21} & b_{22} \\ b_{31} & b_{32} \end{pmatrix}$$

$$= \begin{pmatrix} a_{11}b_{11} + a_{12}b_{21} + a_{13}b_{31} & a_{11}b_{12} + a_{12}b_{22} + a_{13}b_{32} \\ a_{21}b_{11} + a_{22}b_{21} + a_{23}b_{31} & a_{21}b_{12} + a_{22}b_{22} + a_{23}b_{32} \end{pmatrix}$$

If matrix A is of order $(p \times q)$ and matrix B is of order $(q \times r)$ then if $C = AB$ the order of C is $(p \times r)$.

Transposition of a matrix

When the rows of a matrix are interchanged with its columns the matrix is said to b transposed. If the original matrix is denoted by A, its transpose is denoted by A' c A^T.

If $A = \begin{pmatrix} a_{11} & a_{12} & a_{13} \\ a_{21} & a_{22} & a_{23} \end{pmatrix}$ then $A^T = \begin{pmatrix} a_{11} & a_{21} \\ a_{12} & a_{22} \\ a_{13} & a_{23} \end{pmatrix}$

Adjoint of a matrix

If $A = [a_{ij}]$ is any matrix and A_{ij} is the cofactor of a_{ij}, the matrix $[A_{ij}]^T$ is calle the adjoint of A. Thus:

$$A = \begin{pmatrix} a_{11} & a_{12} & \cdots & a_{1n} \\ a_{21} & a_{22} & \cdots & a_{2n} \\ \vdots & \vdots & & \vdots \\ a_{n1} & a_{n2} & \cdots & a_{nn} \end{pmatrix} \qquad \text{adj } A = \begin{pmatrix} A_{11} & A_{21} & \cdots & A_{n1} \\ A_{12} & A_{22} & \cdots & A_{n2} \\ \vdots & \vdots & & \vdots \\ A_{1n} & A_{2n} & \cdots & A_{nn} \end{pmatrix}$$

Singular matrix

A square matrix is singular if the determinant of its coefficients is zero.

The inverse of a matrix

If A is a non-singular matrix of order $(n \times n)$ then its inverse is denoted by A^{-1} suc that $AA^{-1} = I = A^{-1}A$

$$A^{-1} = \frac{\text{adj}(A)}{\Delta} \qquad \Delta = \det(A)$$
$$A_{ij} = \text{cofactor of } a_{ij}$$

If $A = \begin{pmatrix} a_{11} & a_{12} & \cdots & a_{1n} \\ a_{21} & a_{22} & \cdots & a_{2n} \\ \vdots & \vdots & \cdots & \vdots \\ a_{n1} & a_{n2} & \cdots & a_{nn} \end{pmatrix}$ $A^{-1} = \frac{1}{\Delta}\begin{pmatrix} A_{11} & A_{21} & \cdots & A_{n1} \\ A_{12} & A_{22} & \cdots & A_{n2} \\ \vdots & \vdots & \cdots & \vdots \\ A_{1n} & A_{2n} & \cdots & A_{nn} \end{pmatrix}$

Solutions of linear equations

The set of linear equations

$$\begin{aligned} a_{11}x_1 + a_{12}x_2 + \cdots & \quad a_{1n}x_n = b_1 \\ a_{12}x_1 + a_{22}x_2 + \cdots & \quad a_{2n}x_n = b_2 \\ \vdots \qquad \vdots \qquad \cdots & \quad \vdots \quad \vdots \\ a_{1n}x_1 + a_{n2}x_2 + \cdots & \quad a_{nn}x_n = b_n \end{aligned}$$

may be written in matrix form as $Ax = b$

If A is non-singular then the solution is given by $x = A^{-1}b$.

Transformations

The principal transformations are as follows:

(1) To transform $P(x, y)$ into its reflection in the x-axis perform the operation

$$\begin{pmatrix} 1 & 0 \\ 0 & -1 \end{pmatrix}\begin{pmatrix} x \\ y \end{pmatrix}$$

(2) To transform $P(x, y)$ into its reflection in the y-axis perform the operation

$$\begin{pmatrix} -1 & 0 \\ 0 & 1 \end{pmatrix}\begin{pmatrix} x \\ y \end{pmatrix}$$

(3) To transform $P(x, y)$ into its reflection in the origin perform the operation

$$\begin{pmatrix} -1 & 0 \\ 0 & -1 \end{pmatrix}\begin{pmatrix} x \\ y \end{pmatrix}$$

(4) To transform $P(x, y)$ into its reflection in the line $y = x$ perform the operation

$$\begin{pmatrix} 0 & 1 \\ 1 & 0 \end{pmatrix}\begin{pmatrix} x \\ y \end{pmatrix}$$

This transformation reverses the co-ordinates of the point.

Number systems

The number $89\,235$ in the base x means

$$8 \times x^4 + 9 \times x^3 + 2 \times x^2 + 3 \times x^1 + 5 \times x^0$$

Computer programming in BASIC

Constants consist of digits with a decimal point, e.g. 36.84, 191.7.

Variables are specified by a name which consists of either a letter or a letter followed by a digit e.g. A, N7.

Expressions are formed by constants and variables connected by the standard rules of arithmetic.

+	Addition	e.g.	$A + B + F2$
−	Subtraction	e.g.	$B - D$
*	Multiplication	e.g.	$C * L$
/	Division	e.g.	A/F
↑	Exponentiation (raised to a power)	e.g.	$B \uparrow 3$

The order of precedence in an expression is ↑, * or /, + or −.

Round brackets can be used to indicate the order in which operations are to be carried out, e.g. $A + B/C$ is not the same as $(A + B)/C$, also $A/B * C = A/(B * C)$, $A/B * C = \dfrac{A}{B} \times C$, $A/(B * C) = \dfrac{A}{B \times C}$.

Operations of the same procedure are evaluated from left to right.

Line numbers are used to number each statement, these numbers are normally in steps of 10 to allow for the insertion of additional lines. The statements are then executed in ascending order of the statement numbers.

Instruction	Example	Comment
LET	LET P = P + Q/3	Uses the contents of stores P and Q to find the value of $P + Q/3$ and stores the result in store P.
GØ TØ	18 GØ TØ 380	Jumps from line 18 to line 380.
READ	READ A, B	Reads the numbers -17 and 95 from the data and
DATA	DATA -17, 95	stores -17 in store A and 95 in store B.
INPUT	INPUT K, L, T	Numbers are put in at 'run' time and stored in K, L, T.
IF THEN	IF X = 0 THEN 40	Conditional jump, if the content of store X is zero the next line to be worked out will be line 40.
FØR TØ STEP NEXT	70 FØR T = 6 TØ 36 STEP 6 . . . . 140 NEXT T 150	The start of a loop with $T = 6$; T is then increased by 6 each time the NEXT T statement is reached. The programme returns to line 70. When $T > 36$ then line 150 is executed.
DIM	DIM B(10, 5)	This sets up an array of size 10×5 and reserves stores for the $10 \times 5 = 50$ numbers.
DEF FN	DEF FN V(R) = 3*R↑2	V is a function of R, that is $V = 3R^2$.
REM	REM ADD ØN LAST TERM	This is just a remark or comment; it is ignored by the computer.
PRINT	PRINT X, Y	The contents of the stores X and Y are written out.

The following functions are available:

CØS	CØS(X)	The cosine of the angle X (measured in radians) is evaluated.
SIN	SIN(X)	The sine of the angle X (measured in radians) is evaluated.
TAN	TAN(X)	The tangent of the angle X (measured in radians) is evaluated.
EXP	EXP(X)	Evaluates the exponential function e^x.
LØG	LØG(X)	Evaluates the natural logarithm of X.
SQR	SQR(X)	Evaluates the square root of X.
INT	INT(X)	This finds the integer part of X (that is the greatest integer X).
ABS	ABS(X)	This finds the absolute value of X (this is the value of X without its sign).

Other signs used:

=	LET P = A + B	The content of store P is made equal to the sum of the contents of stores A and B.
<	IF X < 0 THEN 60	If the content of store X is less than 0 then go to line 60.
>	IF X > 0 THEN 90	If the content of store X is greater than 0 then go to line 90.
<=	IF X <= 0 THEN 100	If the content of store X is less than or equal to 0 then go to line 100.
>=	IF X >= 0 THEN 50	If the content of store X is greater than or equal to 0 then go to line 50.
<>	IF X <> 0 THEN 80	If the content of store X is not equal to 0 then go to line 80.

Commonly used constants

Constant	Numerical value	Logarithm	Constant	Numerical value	Logarithm
π	3.141593	0.4972	$1/\pi$	0.318310	$\bar{1}.5029$
2π	6.283185	0.7982	$\sqrt{\pi}$	1.772454	0.2486
$\pi/4$	0.785398	$\bar{1}.8951$	e	2.71828	0.4343
π^2	9.869604	0.9943	g	9.81	0.9917

Atmospheric pressure $= 760\,\text{mm Hg} = 1013\,\text{mb}$ 1 bar $= 10^2 \text{KN/m}^2$
Diameter of the earth $= 12\,750\,\text{km}$ at the equator and $12\,710\,\text{km}$ at the poles.
Average radius of the earth $= 6371\,\text{km}$
Speed of rotation of the earth $= 1670\,\text{km/h}$

Calculator check

Not all calculators have the same logic and the keys and stores do not always work in the same way. Therefore before starting to perform strings of calculations it pays to check that the calculator is working correctly. The following can be used to check the logic of the calculator.

$67.84 + 91.92 + 71.85 = 231.61$

$66.32 - 19.85 = 46.47$

$88.56 - 13.84 + 24.31 = 99.03$

$77.3 \times 64.8 = 5009.04$

$91.76 \times 3.84 + 817.52 = 1169.8784$

$(7.85 + 3.91) \times 83.64 = 983.6064$

$91.3 \times 43.2 \times 68.0 = 268\,202.88$

$\dfrac{91.76}{1.85} = 49.6$

$\dfrac{81.32 \times 14.63}{76.51} = 15.549\,752$

$\dfrac{84.3}{91.2} + \dfrac{76.51}{3.84} = 20.848\,821$

for calculators without a memory this may be calculated thus

$\left(\dfrac{84.3 \times 3.84}{91.2} + 76.51 \right) \div 3.84 = 20.848\,821$

$\dfrac{816.1}{94.3} - \dfrac{36.2}{14.7} = 6.191\,709\,8$

for calculators without a memory this may be calculated thus

$\left(\dfrac{-36.2 \times 94.3}{14.7} + 816.1 \right) \div 94.3 = 6.191\,709\,8$

$17.62 - \dfrac{8.54}{3.61} = 15.254\,35$

Logarithms

	0	1	2	3	4	5	6	7	8	9
10	0000	0043	0086	0128	0170	0212	0253	0294	0334	0374
11	0414	0453	0492	0531	0569	0607	0645	0682	0719	0755
12	0792	0828	0864	0899	0934	0969	1004	1038	1072	1106
13	1139	1173	1206	1239	1271	1303	1335	1367	1399	1430
14	1461	1492	1523	1553	1584	1614	1644	1673	1703	1732
15	1761	1790	1818	1847	1875	1903	1931	1959	1987	2014
16	2041	2068	2095	2122	2148	2175	2201	2227	2253	2279
17	2304	2330	2355	2380	2405	2430	2455	2480	2504	2529
18	2553	2577	2601	2625	2648	2672	2695	2718	2742	2765
19	2788	2810	2833	2856	2878	2900	2923	2945	2967	2989
20	3010	3032	3054	3075	3096	3118	3139	3160	3181	3201
21	3222	3243	3263	3284	3304	3324	3345	3365	3385	3404
22	3424	3444	3464	3483	3502	3522	3541	3560	3579	3598
23	3617	3636	3655	3674	3692	3711	3729	3747	3766	3784
24	3802	3820	3838	3856	3874	3892	3909	3927	3945	3962
25	3979	3997	4014	4031	4048	4065	4082	4099	4116	4133
26	4150	4166	4183	4200	4216	4232	4249	4265	4281	4298
27	4314	4330	4346	4362	4378	4393	4409	4425	4440	4456
28	4472	4487	4502	4518	4533	4548	4564	4579	4594	4609
29	4624	4639	4654	4669	4683	4698	4713	4728	4742	4757
30	4771	4786	4800	4814	4829	4843	4857	4871	4886	4900
31	4914	4928	4942	4955	4969	4983	4997	5011	5024	5038
32	5051	5065	5079	5092	5105	5119	5132	5145	5159	5172
33	5185	5198	5211	5224	5237	5250	5263	5276	5289	5302
34	5315	5328	5340	5353	5366	5378	5391	5403	5416	5428
35	5441	5453	5465	5478	5490	5502	5514	5527	5539	5551
36	5563	5575	5587	5599	5611	5623	5635	5647	5658	5670
37	5682	5694	5705	5717	5729	5740	5752	5763	5775	5786
38	5798	5809	5821	5832	5843	5855	5866	5877	5888	5899
39	5911	5922	5933	5944	5955	5966	5977	5988	5999	6010
40	6021	6031	6042	6053	6064	6075	6085	6096	6107	6117
41	6128	6138	6149	6160	6170	6180	6191	6201	6212	6222
42	6232	6243	6253	6263	6274	6284	6294	6304	6314	6325
43	6335	6345	6355	6365	6375	6385	6395	6405	6415	6425
44	6435	6444	6454	6464	6474	6484	6493	6503	6513	6522
45	6532	6542	6551	6561	6571	6580	6590	6599	6609	6618
46	6628	6637	6646	6656	6665	6675	6684	6693	6702	6712
47	6721	6730	6739	6749	6758	6767	6776	6785	6794	6803
48	6812	6821	6830	6839	6848	6857	6866	6875	6884	6893
49	6902	6911	6920	6928	6937	6946	6955	6964	6972	6981

Mean differences:

1	2	3	4	5	6	7	8	9
4	8	13	17	21	25	30	34	38
4	8	12	16	20	24	28	32	36
4	8	12	15	19	23	27	31	35
4	7	11	15	18	22	26	30	33
4	7	11	14	18	21	25	28	32
3	7	10	14	17	20	24	27	31
3	7	10	13	16	20	23	26	30
3	6	9	13	16	19	22	25	28
3	6	9	12	15	18	21	24	27
3	6	9	12	15	18	21	24	27
3	6	9	11	14	17	20	23	26
3	6	8	11	14	17	19	22	25
3	5	8	11	13	16	19	21	24
3	5	8	10	13	16	18	21	23
3	5	8	10	13	15	18	20	23
2	5	7	10	12	15	17	20	22
2	5	7	9	12	14	16	19	21
2	5	7	9	11	14	16	18	21
2	5	7	9	11	14	16	18	20
2	4	7	9	11	13	15	18	20
2	4	6	8	11	13	15	17	19
2	4	6	8	10	12	14	16	18
2	4	6	8	10	12	14	15	17
2	4	6	7	9	11	13	15	17
2	4	5	7	9	11	12	14	16
2	3	5	7	9	10	12	14	15
2	3	5	7	8	10	11	13	15
2	3	5	6	8	9	11	13	14
2	3	5	6	8	9	11	12	14
1	3	4	6	7	9	10	12	13
1	3	4	6	7	9	10	11	13
1	3	4	6	7	8	10	11	12
1	3	4	5	7	8	9	11	12
1	3	4	5	6	8	9	10	12
1	3	4	5	6	8	9	10	11
1	2	4	5	6	7	9	10	11
1	2	4	5	6	7	8	10	11
1	2	3	5	6	7	8	9	10
1	2	3	5	6	7	8	9	10
1	2	3	4	5	7	8	9	10
1	2	3	4	5	6	8	9	10
1	2	3	4	5	6	7	8	9
1	2	3	4	5	6	7	8	9
1	2	3	4	5	6	7	8	9
1	2	3	4	5	6	7	8	9
1	2	3	4	5	6	7	8	9
1	2	3	4	5	6	7	7	8
1	2	3	4	5	5	6	7	8
1	2	3	4	4	5	6	7	8
1	2	3	4	4	5	6	7	8

$$\log(3.546 \times 8.714) = \log 3.546 + \log 8.714$$
$$= 0.5497 + 0.9402 = 1.4899$$

by taking antilog of 1.4899
$$3.546 \times 8.714 = 30.89$$

$$\log\left(\frac{7.362}{5.169}\right) = \log 7.362 - \log 5.169$$
$$= 0.8670 - 0.7134 = 0.1536$$

by taking antilog of 0.1536
$$\left(\frac{7.362}{5.169}\right) = 1.424$$

The characteristic for a number greater than 1 is found by subtracting 1 from the number of figures to the left of the decimal point.

$$\log 384.2 = 2.5845 \qquad \log 59\,860 = 4.7771$$

Logarithms

	0	1	2	3	4	5	6	7	8	9	1	2	3	4	5	6	7	8	9
50	6990	6998	7007	7016	7024	7033	7042	7050	7059	7067	1	2	3	3	4	5	6	7	8
51	7076	7084	7093	7101	7110	7118	7126	7135	7143	7152	1	2	3	3	4	5	6	7	8
52	7160	7168	7177	7185	7193	7202	7210	7218	7226	7235	1	2	2	3	4	5	6	7	7
53	7243	7251	7259	7267	7275	7284	7292	7300	7308	7316	1	2	2	3	4	5	6	6	7
54	7324	7332	7340	7348	7356	7364	7372	7380	7388	7396	1	2	2	3	4	5	6	6	7
55	7404	7412	7419	7427	7435	7443	7451	7459	7466	7474	1	2	2	3	4	5	5	6	7
56	7482	7490	7497	7505	7513	7520	7528	7536	7543	7551	1	2	2	3	4	5	5	6	7
57	7559	7566	7574	7582	7589	7597	7604	7612	7619	7627	1	2	2	3	4	5	5	6	7
58	7634	7642	7649	7657	7664	7672	7679	7686	7694	7701	1	1	2	3	4	4	5	6	7
59	7709	7716	7723	7731	7738	7745	7752	7760	7767	7774	1	1	2	3	4	4	5	6	7
60	7782	7789	7796	7803	7810	7818	7825	7832	7839	7846	1	1	2	3	4	4	5	6	6
61	7853	7860	7868	7875	7882	7889	7896	7903	7910	7917	1	1	2	3	4	4	5	6	6
62	7924	7931	7938	7945	7952	7959	7966	7973	7980	7987	1	1	2	3	3	4	5	6	6
63	7993	8000	8007	8014	8021	8028	8035	8041	8048	8055	1	1	2	3	3	4	5	5	6
64	8062	8069	8075	8082	8089	8096	8102	8109	8116	8122	1	1	2	3	3	4	5	5	6
65	8129	8136	8142	8149	8156	8162	8169	8176	8182	8189	1	1	2	3	3	4	5	5	6
66	8195	8202	8209	8215	8222	8228	8235	8241	8248	8254	1	1	2	3	3	4	5	5	6
67	8261	8267	8274	8280	8287	8293	8299	8306	8312	8319	1	1	2	3	3	4	5	5	6
68	8325	8331	8338	8344	8351	8357	8363	8370	8376	8382	1	1	2	3	3	4	4	5	6
69	8388	8395	8401	8407	8414	8420	8426	8432	8439	8445	1	1	2	2	3	4	4	5	6
70	8451	8457	8463	8470	8476	8482	8488	8494	8500	8506	1	1	2	2	3	4	4	5	6
71	8513	8519	8525	8531	8537	8543	8549	8555	8561	8567	1	1	2	2	3	4	4	5	5
72	8573	8579	8585	8591	8597	8603	8609	8615	8621	8627	1	1	2	2	3	4	4	5	5
73	8633	8639	8645	8651	8657	8663	8669	8675	8681	8686	1	1	2	2	3	4	4	5	5
74	8692	8698	8704	8710	8716	8722	8727	8733	8739	8745	1	1	2	2	3	4	4	5	5
75	8751	8756	8762	8768	8774	8779	8785	8791	8797	8802	1	1	2	2	3	3	4	5	5
76	8808	8814	8820	8825	8831	8837	8842	8848	8854	8859	1	1	2	2	3	3	4	5	5
77	8865	8871	8876	8882	8887	8893	8899	8904	8910	8915	1	1	2	2	3	3	4	4	5
78	8921	8927	8932	8938	8943	8949	8954	8960	8965	8971	1	1	2	2	3	3	4	4	5
79	8976	8982	8987	8993	8998	9004	9009	9015	9020	9025	1	1	2	2	3	3	4	4	5
80	9031	9036	9042	9047	9053	9058	9063	9069	9074	9079	1	1	2	2	3	3	4	4	5
81	9085	9090	9096	9101	9106	9112	9117	9122	9128	9133	1	1	2	2	3	3	4	4	5
82	9138	9143	9149	9154	9159	9165	9170	9175	9180	9186	1	1	2	2	3	3	4	4	5
83	9191	9196	9201	9206	9212	9217	9222	9227	9232	9238	1	1	2	2	3	3	4	4	5
84	9243	9248	9253	9258	9263	9269	9274	9279	9284	9289	1	1	2	2	3	3	4	4	5
85	9294	9299	9304	9309	9315	9320	9325	9330	9335	9340	1	1	2	2	3	3	4	4	5
86	9345	9350	9355	9360	9365	9370	9375	9380	9385	9390	1	1	2	2	3	3	4	4	5
87	9395	9400	9405	9410	9415	9420	9425	9430	9435	9440	0	1	1	2	2	3	3	4	4
88	9445	9450	9455	9460	9465	9469	9474	9479	9484	9489	0	1	1	2	2	3	3	4	4
89	9494	9499	9504	9509	9513	9518	9523	9528	9533	9538	0	1	1	2	2	3	3	4	4
90	9542	9547	9552	9557	9562	9566	9571	9576	9581	9586	0	1	1	2	2	3	3	4	4
91	9590	9595	9600	9605	9609	9614	9619	9624	9628	9633	0	1	1	2	2	3	3	4	4
92	9638	9643	9647	9652	9657	9661	9666	9671	9675	9680	0	1	1	2	2	3	3	4	4
93	9685	9689	9694	9699	9703	9708	9713	9717	9722	9727	0	1	1	2	2	3	3	4	4
94	9731	9736	9741	9745	9750	9754	9759	9763	9768	9773	0	1	1	2	2	3	3	4	4
95	9777	9782	9786	9791	9795	9800	9805	9809	9814	9818	0	1	1	2	2	3	3	4	4
96	9823	9827	9832	9836	9841	9845	9850	9854	9859	9863	0	1	1	2	2	3	3	4	4
97	9868	9872	9877	9881	9886	9890	9894	9899	9903	9908	0	1	1	2	2	3	3	4	4
98	9912	9917	9921	9926	9930	9934	9939	9943	9948	9952	0	1	1	2	2	3	3	4	4
99	9956	9961	9965	9969	9974	9978	9983	9987	9991	9996	0	1	1	2	2	3	3	3	4

$$\log(2.534)^3 = 3 \times \log 2.534$$
$$= 3 \times 0.4038 = 1.2114$$
by taking antilog of 1.2114
$$(2.534)^3 = 16.28$$

$$\log {}^5\sqrt{8.176} = \tfrac{1}{5} \times \log 8.176$$
$$= \tfrac{1}{5} \times 0.9125 = 0.1825$$
by taking antilog of 0.1825
$${}^5\sqrt{8.176} = 1.523$$

The negative characteristic for a number between 0 and 1 is found by adding 1 to the number of zeros following the decimal point.
$$\log 0.0578 = \bar{2}.7619 \qquad \log 0.000\,432 = \bar{4}.6355$$

Antilogarithms

	0	1	2	3	4	5	6	7	8	9	1	2	3	4	5	6	7	8	9
0.00	1000	1002	1005	1007	1009	1012	1014	1016	1019	1021	0	0	1	1	1	1	2	2	2
0.01	1023	1026	1028	1030	1033	1035	1038	1040	1042	1045	0	0	1	1	1	1	2	2	2
0.02	1047	1050	1052	1054	1057	1059	1062	1064	1067	1069	0	0	1	1	1	1	2	2	2
0.03	1072	1074	1076	1079	1081	1084	1086	1089	1091	1094	0	0	1	1	1	1	2	2	2
0.04	1096	1099	1102	1104	1107	1109	1112	1114	1117	1119	0	1	1	1	1	2	2	2	2
0.05	1122	1125	1127	1130	1132	1135	1138	1140	1143	1146	0	1	1	1	1	2	2	2	2
0.06	1148	1151	1153	1156	1159	1161	1164	1167	1169	1172	0	1	1	1	1	2	2	2	2
0.07	1175	1178	1180	1183	1186	1189	1191	1194	1197	1199	0	1	1	1	1	2	2	2	2
0.08	1202	1205	1208	1211	1213	1216	1219	1222	1225	1227	0	1	1	1	1	2	2	2	3
0.09	1230	1233	1236	1239	1242	1245	1247	1250	1253	1256	0	1	1	1	1	2	2	2	3
0.10	1259	1262	1265	1268	1271	1274	1276	1279	1282	1285	0	1	1	1	1	2	2	2	3
0.11	1288	1291	1294	1297	1300	1303	1306	1309	1312	1315	0	1	1	1	2	2	2	2	3
0.12	1318	1321	1324	1327	1330	1334	1337	1340	1343	1346	0	1	1	1	2	2	2	3	3
0.13	1349	1352	1355	1358	1361	1365	1368	1371	1374	1377	0	1	1	1	2	2	2	3	3
0.14	1380	1384	1387	1390	1393	1396	1400	1403	1406	1409	0	1	1	1	2	2	2	3	3
0.15	1413	1416	1419	1422	1426	1429	1432	1435	1439	1442	0	1	1	1	2	2	2	3	3
0.16	1445	1449	1452	1455	1459	1462	1466	1469	1472	1476	0	1	1	1	2	2	2	3	3
0.17	1479	1483	1486	1489	1493	1496	1500	1503	1507	1510	0	1	1	1	2	2	2	3	3
0.18	1514	1517	1521	1524	1528	1531	1535	1538	1542	1545	0	1	1	1	2	2	2	3	3
0.19	1549	1552	1556	1560	1563	1567	1570	1574	1578	1581	0	1	1	1	2	2	3	3	3
0.20	1585	1589	1592	1596	1600	1603	1607	1611	1614	1618	0	1	1	1	2	2	3	3	3
0.21	1622	1626	1629	1633	1637	1641	1644	1648	1652	1656	0	1	1	2	2	2	3	3	3
0.22	1660	1663	1667	1671	1675	1679	1683	1687	1690	1694	0	1	1	2	2	2	3	3	3
0.23	1698	1702	1706	1710	1714	1718	1722	1726	1730	1734	0	1	1	2	2	2	3	3	4
0.24	1738	1742	1746	1750	1754	1758	1762	1766	1770	1774	0	1	1	2	2	2	3	3	4
0.25	1778	1782	1786	1791	1795	1799	1803	1807	1811	1816	0	1	1	2	2	2	3	3	4
0.26	1820	1824	1828	1832	1837	1841	1845	1849	1854	1858	0	1	1	2	2	3	3	3	4
0.27	1862	1866	1871	1875	1879	1884	1888	1892	1897	1901	0	1	1	2	2	3	3	3	4
0.28	1905	1910	1914	1919	1923	1928	1932	1936	1941	1945	0	1	1	2	2	3	3	4	4
0.29	1950	1954	1959	1963	1968	1972	1977	1982	1986	1991	0	1	1	2	2	3	3	4	4
0.30	1995	2000	2004	2009	2014	2018	2023	2028	2032	2037	0	1	1	2	2	3	3	4	4
0.31	2042	2046	2051	2056	2061	2065	2070	2075	2080	2084	0	1	1	2	2	3	3	4	4
0.32	2089	2094	2099	2104	2109	2113	2118	2123	2128	2133	0	1	1	2	2	3	3	4	4
0.33	2138	2143	2148	2153	2158	2163	2168	2173	2178	2183	0	1	1	2	2	3	3	4	4
0.34	2188	2193	2198	2203	2208	2213	2218	2223	2228	2234	1	1	2	2	3	3	4	4	5
0.35	2239	2244	2249	2254	2259	2265	2270	2275	2280	2286	1	1	2	2	3	3	4	4	5
0.36	2291	2296	2301	2307	2312	2317	2323	2328	2333	2339	1	1	2	2	3	3	4	4	5
0.37	2344	2350	2355	2360	2366	2371	2377	2382	2388	2393	1	1	2	2	3	3	4	4	5
0.38	2399	2404	2410	2415	2421	2427	2432	2438	2443	2449	1	1	2	2	3	3	4	4	5
0.39	2455	2460	2466	2472	2477	2483	2489	2495	2500	2506	1	1	2	2	3	3	4	5	5
0.40	2512	2518	2523	2529	2535	2541	2547	2553	2559	2564	1	1	2	2	3	4	4	5	5
0.41	2570	2576	2582	2588	2594	2600	2606	2612	2618	2624	1	1	2	2	3	4	4	5	5
0.42	2630	2636	2642	2649	2655	2661	2667	2673	2679	2685	1	1	2	2	3	4	4	5	6
0.43	2692	2698	2704	2710	2716	2723	2729	2735	2742	2748	1	1	2	2	3	4	4	5	6
0.44	2754	2761	2767	2773	2780	2786	2793	2799	2805	2812	1	1	2	3	3	4	4	5	6
0.45	2818	2825	2831	2838	2844	2851	2858	2864	2871	2877	1	1	2	3	3	4	5	5	6
0.46	2884	2891	2897	2904	2911	2917	2924	2931	2938	2944	1	1	2	3	3	4	5	5	6
0.47	2951	2958	2965	2972	2979	2985	2992	2999	3006	3013	1	1	2	3	3	4	5	5	6
0.48	3020	3027	3034	3041	3048	3055	3062	3069	3076	3083	1	1	2	3	4	4	5	6	6
0.49	3090	3097	3105	3112	3119	3126	3133	3141	3148	3155	1	1	2	3	4	4	5	6	6

Only the mantissa (decimal part) of a logarithm is used when using the anti-log tables.

To find the number whose log is 2.2004.

Using the mantissa .2004 the number corresponding is 1586.

Since the characteristic is 2 the number must have three figures to the left of the decimal point. The number is 158.6. (Note that log 158.6 = 2.2004.)

Antilogarithms

	0	1	2	3	4	5	6	7	8	9	1	2	3	4	5	6	7	8	9
0.50	3162	3170	3177	3184	3192	3199	3206	3214	3221	3228	1	1	2	3	4	4	5	6	7
0.51	3236	3243	3251	3258	3266	3273	3281	3289	3296	3304	1	2	2	3	4	5	5	6	7
0.52	3311	3319	3327	3334	3342	3350	3357	3365	3373	3381	1	2	2	3	4	5	5	6	7
0.53	3388	3396	3404	3412	3420	3428	3436	3443	3451	3459	1	2	2	3	4	5	6	6	7
0.54	3467	3475	3483	3491	3499	3508	3516	3524	3532	3540	1	2	2	3	4	5	6	6	7
0.55	3548	3556	3565	3573	3581	3589	3597	3606	3614	3622	1	2	2	3	4	5	6	7	7
0.56	3631	3639	3648	3656	3664	3673	3681	3690	3698	3707	1	2	3	3	4	5	6	7	8
0.57	3715	3724	3733	3741	3750	3758	3767	3776	3784	3793	1	2	3	3	4	5	6	7	8
0.58	3802	3811	3819	3828	3837	3846	3855	3864	3873	3882	1	2	3	4	4	5	6	7	8
0.59	3890	3899	3908	3917	3926	3936	3945	3954	3963	3972	1	2	3	4	5	5	6	7	8
0.60	3981	3990	3999	4009	4018	4027	4036	4046	4055	4064	1	2	3	4	5	6	6	7	8
0.61	4074	4083	4093	4102	4111	4121	4130	4140	4150	4159	1	2	3	4	5	6	7	8	9
0.62	4169	4178	4188	4198	4207	4217	4227	4236	4246	4256	1	2	3	4	5	6	7	8	9
0.63	4266	4276	4285	4295	4305	4315	4325	4335	4345	4355	1	2	3	4	5	6	7	8	9
0.64	4365	4375	4385	4395	4406	4416	4426	4436	4446	4457	1	2	3	4	5	6	7	8	9
0.65	4467	4477	4487	4498	4508	4519	4529	4539	4550	4560	1	2	3	4	5	6	7	8	9
0.66	4571	4581	4592	4603	4613	4624	4634	4645	4656	4667	1	2	3	4	5	6	7	9	10
0.67	4677	4688	4699	4710	4721	4732	4742	4753	4764	4775	1	2	3	4	5	7	8	9	10
0.68	4786	4797	4808	4819	4831	4842	4853	4864	4875	4887	1	2	3	4	6	7	8	9	10
0.69	4893	4909	4920	4932	4943	4955	4966	4977	4989	5000	1	2	3	5	6	7	8	9	10
0.70	5012	5023	5035	5047	5058	5070	5082	5093	5105	5117	1	2	4	5	6	7	8	9	11
0.71	5129	5140	5152	5164	5176	5188	5200	5212	5224	5236	1	2	4	5	6	7	8	10	11
0.72	5248	5260	5272	5284	5297	5309	5321	5333	5336	5358	1	2	4	5	6	7	9	10	11
0.73	5370	5383	5395	5408	5420	5433	5445	5458	5470	5483	1	3	4	5	6	8	9	10	11
0.74	5495	5508	5521	5534	5546	5559	5572	5585	5598	5610	1	3	4	5	6	8	9	10	12
0.75	5623	5636	5649	5662	5675	5689	5702	5715	5728	5741	1	3	4	5	7	8	9	10	12
0.76	5754	5768	5781	5794	5808	5821	5834	5848	5861	5875	1	3	4	5	7	8	9	11	12
0.77	5888	5902	5916	5929	5943	5957	5970	5984	5998	6012	1	3	4	5	7	8	10	11	12
0.78	6026	6039	6053	6067	6081	6095	6109	6124	6138	6152	1	3	4	6	7	8	10	11	13
0.79	6166	6180	6194	6209	6223	6237	6252	6266	6281	6295	1	3	4	6	7	9	10	11	13
0.80	6310	6324	6339	6353	6368	6383	6397	6412	6427	6442	1	3	4	6	7	9	10	12	13
0.81	6457	6471	6486	6501	6516	6531	6546	6561	6577	6592	2	3	5	6	8	9	11	12	14
0.82	6607	6622	6637	6653	6668	6683	6699	6714	6730	6745	2	3	5	6	8	9	11	12	14
0.83	6761	6776	6792	6808	6823	6839	6855	6871	6887	6902	2	3	5	6	8	9	11	13	14
0.84	6918	6934	6950	6966	6982	6998	7015	7031	7047	7063	2	3	5	6	8	10	11	13	15
0.85	7079	7096	7112	7129	7145	7161	7178	7194	7211	7228	2	3	5	7	8	10	12	13	15
0.86	7244	7261	7278	7295	7311	7328	7345	7362	7379	7396	2	3	5	7	8	10	12	13	15
0.87	7413	7430	7447	7464	7482	7499	7516	7534	7551	7568	2	3	5	7	9	10	12	14	16
0.88	7586	7603	7621	7638	7656	7674	7691	7709	7727	7745	2	4	5	7	9	11	12	14	16
0.89	7762	7780	7798	7816	7834	7852	7870	7889	7907	7925	2	4	5	7	9	11	13	14	16
0.90	7943	7962	7980	7998	8017	8035	8054	8072	8091	8110	2	4	6	7	9	11	13	15	17
0.91	8128	8147	8166	8185	8204	8222	8241	8260	8279	8299	2	4	6	8	9	11	13	15	17
0.92	8318	8337	8356	8375	8395	8414	8433	8453	8472	8492	2	4	6	8	10	12	14	15	17
0.93	8511	8531	8551	8570	8590	8610	8630	8650	8670	8690	2	4	6	8	10	12	14	16	18
0.94	8710	8730	8750	8770	8790	8810	8831	8851	8872	8892	2	4	6	8	10	12	14	16	18
0.95	8913	8933	8954	8974	8995	9016	9036	9057	9078	9099	2	4	6	8	10	12	15	17	19
0.96	9120	9141	9162	9183	9204	9226	9247	9268	9290	9311	2	4	6	8	11	13	15	17	19
0.97	9333	9354	9376	9397	9419	9441	9462	9484	9506	9528	2	4	7	9	11	13	15	17	20
0.98	9550	9572	9594	9616	9638	9661	9683	9705	9727	9750	2	4	7	9	11	13	16	18	20
0.99	9772	9795	9817	9840	9863	9886	9908	9931	9954	9977	2	5	7	9	11	14	16	18	20

To find the number whose log is 3.8178.
Using the mantissa .8178 the number corresponding is 6573.
Since the characteristic is $\bar{3}$ there must be two zeros following the decimal point. The number is 0.006 573.
(Note that log 0.006 573 is $\bar{3}$.8178.)

Logarithms of Sines

°	0' 0.0°	6' 0.1°	12' 0.2°	18' 0.3°	24' 0.4°	30' 0.5°	36' 0.6°	42' 0.7°	48' 0.8°	54' 0.9°	1'	2'	3'	4'	5'
0	−∞	$\overline{3}.2419$	$\overline{3}.5429$	$\overline{3}.7190$	$\overline{3}.8439$	$\overline{3}.9408$	$\overline{2}.0200$	$\overline{2}.0870$	$\overline{2}.1450$	$\overline{2}.1961$		Differences			
1	$\overline{2}.2419$	$\overline{2}.2832$	$\overline{2}.3210$	$\overline{2}.3558$	$\overline{2}.3880$	$\overline{2}.4179$	$\overline{2}.4459$	$\overline{2}.4723$	$\overline{2}.4971$	$\overline{2}.5206$		untrustworthy			
2	$\overline{2}.5428$	$\overline{2}.5640$	$\overline{2}.5842$	$\overline{2}.6035$	$\overline{2}.6220$	$\overline{2}.6397$	$\overline{2}.6567$	$\overline{2}.6731$	$\overline{2}.6889$	$\overline{2}.7041$		here			
3	$\overline{2}.7188$	$\overline{2}.7330$	$\overline{2}.7468$	$\overline{2}.7602$	$\overline{2}.7731$	$\overline{2}.7857$	$\overline{2}.7979$	$\overline{2}.8098$	$\overline{2}.8213$	$\overline{2}.8326$					
4	$\overline{2}.8436$	$\overline{2}.8543$	$\overline{2}.8647$	$\overline{2}.8749$	$\overline{2}.8849$	$\overline{2}.8946$	$\overline{2}.9042$	$\overline{2}.9135$	$\overline{2}.9226$	$\overline{2}.9315$	16	32	48	64	80
5	$\overline{2}.9403$	$\overline{2}.9489$	$\overline{2}.9573$	$\overline{2}.9655$	$\overline{2}.9736$	$\overline{2}.9816$	$\overline{2}.9894$	$\overline{2}.9970$	$\overline{1}.0046$	$\overline{1}.0120$	13	26	39	52	65
6	$\overline{1}.0192$	$\overline{1}.0264$	$\overline{1}.0334$	$\overline{1}.0403$	$\overline{1}.0472$	$\overline{1}.0539$	$\overline{1}.0605$	$\overline{1}.0670$	$\overline{1}.0734$	$\overline{1}.0797$	11	22	33	44	55
7	$\overline{1}.0859$	$\overline{1}.0920$	$\overline{1}.0981$	$\overline{1}.1040$	$\overline{1}.1099$	$\overline{1}.1157$	$\overline{1}.1214$	$\overline{1}.1271$	$\overline{1}.1326$	$\overline{1}.1381$	10	19	29	38	48
8	$\overline{1}.1436$	$\overline{1}.1489$	$\overline{1}.1542$	$\overline{1}.1594$	$\overline{1}.1646$	$\overline{1}.1697$	$\overline{1}.1747$	$\overline{1}.1797$	$\overline{1}.1847$	$\overline{1}.1895$	8	17	25	34	42
9	$\overline{1}.1943$	$\overline{1}.1991$	$\overline{1}.2038$	$\overline{1}.2085$	$\overline{1}.2131$	$\overline{1}.2176$	$\overline{1}.2221$	$\overline{1}.2266$	$\overline{1}.2310$	$\overline{1}.2353$	8	15	23	30	38
10	$\overline{1}.2397$	$\overline{1}.2439$	$\overline{1}.2482$	$\overline{1}.2524$	$\overline{1}.2565$	$\overline{1}.2606$	$\overline{1}.2647$	$\overline{1}.2687$	$\overline{1}.2727$	$\overline{1}.2767$	7	14	20	27	34
11	$\overline{1}.2806$	$\overline{1}.2845$	$\overline{1}.2883$	$\overline{1}.2921$	$\overline{1}.2959$	$\overline{1}.2997$	$\overline{1}.3034$	$\overline{1}.3070$	$\overline{1}.3107$	$\overline{1}.3143$	6	12	19	25	31
12	$\overline{1}.3179$	$\overline{1}.3214$	$\overline{1}.3250$	$\overline{1}.3284$	$\overline{1}.3319$	$\overline{1}.3353$	$\overline{1}.3387$	$\overline{1}.3421$	$\overline{1}.3455$	$\overline{1}.3488$	6	11	17	23	28
13	$\overline{1}.3521$	$\overline{1}.3554$	$\overline{1}.3586$	$\overline{1}.3618$	$\overline{1}.3650$	$\overline{1}.3682$	$\overline{1}.3713$	$\overline{1}.3745$	$\overline{1}.3775$	$\overline{1}.3806$	5	11	16	21	26
14	$\overline{1}.3837$	$\overline{1}.3867$	$\overline{1}.3897$	$\overline{1}.3927$	$\overline{1}.3957$	$\overline{1}.3986$	$\overline{1}.4015$	$\overline{1}.4044$	$\overline{1}.4073$	$\overline{1}.4102$	5	10	15	20	24
15	$\overline{1}.4130$	$\overline{1}.4158$	$\overline{1}.4186$	$\overline{1}.4214$	$\overline{1}.4242$	$\overline{1}.4269$	$\overline{1}.4296$	$\overline{1}.4323$	$\overline{1}.4350$	$\overline{1}.4377$	5	9	14	18	23
16	$\overline{1}.4403$	$\overline{1}.4430$	$\overline{1}.4456$	$\overline{1}.4482$	$\overline{1}.4508$	$\overline{1}.4533$	$\overline{1}.4559$	$\overline{1}.4584$	$\overline{1}.4609$	$\overline{1}.4634$	4	9	13	17	21
17	$\overline{1}.4659$	$\overline{1}.4684$	$\overline{1}.4709$	$\overline{1}.4733$	$\overline{1}.4757$	$\overline{1}.4781$	$\overline{1}.4805$	$\overline{1}.4829$	$\overline{1}.4853$	$\overline{1}.4876$	4	8	12	16	20
18	$\overline{1}.4900$	$\overline{1}.4923$	$\overline{1}.4946$	$\overline{1}.4969$	$\overline{1}.4992$	$\overline{1}.5015$	$\overline{1}.5037$	$\overline{1}.5060$	$\overline{1}.5082$	$\overline{1}.5104$	4	8	11	15	19
19	$\overline{1}.5126$	$\overline{1}.5148$	$\overline{1}.5170$	$\overline{1}.5192$	$\overline{1}.5213$	$\overline{1}.5235$	$\overline{1}.5256$	$\overline{1}.5278$	$\overline{1}.5299$	$\overline{1}.5320$	4	7	11	14	18
20	$\overline{1}.5341$	$\overline{1}.5361$	$\overline{1}.5382$	$\overline{1}.5402$	$\overline{1}.5423$	$\overline{1}.5443$	$\overline{1}.5463$	$\overline{1}.5484$	$\overline{1}.5504$	$\overline{1}.5523$	3	7	10	13	17
21	$\overline{1}.5543$	$\overline{1}.5563$	$\overline{1}.5583$	$\overline{1}.5602$	$\overline{1}.5621$	$\overline{1}.5641$	$\overline{1}.5660$	$\overline{1}.5679$	$\overline{1}.5698$	$\overline{1}.5717$	3	6	10	13	16
22	$\overline{1}.5736$	$\overline{1}.5754$	$\overline{1}.5773$	$\overline{1}.5792$	$\overline{1}.5810$	$\overline{1}.5828$	$\overline{1}.5847$	$\overline{1}.5865$	$\overline{1}.5883$	$\overline{1}.5901$	3	6	9	12	15
23	$\overline{1}.5919$	$\overline{1}.5937$	$\overline{1}.5954$	$\overline{1}.5972$	$\overline{1}.5990$	$\overline{1}.6007$	$\overline{1}.6024$	$\overline{1}.6042$	$\overline{1}.6059$	$\overline{1}.6076$	3	6	9	12	14
24	$\overline{1}.6093$	$\overline{1}.6110$	$\overline{1}.6127$	$\overline{1}.6144$	$\overline{1}.6161$	$\overline{1}.6177$	$\overline{1}.6194$	$\overline{1}.6210$	$\overline{1}.6227$	$\overline{1}.6243$	3	6	8	11	14
25	$\overline{1}.6259$	$\overline{1}.6276$	$\overline{1}.6292$	$\overline{1}.6308$	$\overline{1}.6324$	$\overline{1}.6340$	$\overline{1}.6356$	$\overline{1}.6371$	$\overline{1}.6387$	$\overline{1}.6403$	3	5	8	11	13
26	$\overline{1}.6418$	$\overline{1}.6434$	$\overline{1}.6449$	$\overline{1}.6465$	$\overline{1}.6480$	$\overline{1}.6495$	$\overline{1}.6510$	$\overline{1}.6526$	$\overline{1}.6541$	$\overline{1}.6556$	3	5	8	10	13
27	$\overline{1}.6570$	$\overline{1}.6585$	$\overline{1}.6600$	$\overline{1}.6615$	$\overline{1}.6629$	$\overline{1}.6644$	$\overline{1}.6659$	$\overline{1}.6673$	$\overline{1}.6687$	$\overline{1}.6702$	2	5	7	10	12
28	$\overline{1}.6716$	$\overline{1}.6730$	$\overline{1}.6744$	$\overline{1}.6759$	$\overline{1}.6773$	$\overline{1}.6787$	$\overline{1}.6801$	$\overline{1}.6814$	$\overline{1}.6828$	$\overline{1}.6842$	2	5	7	9	12
29	$\overline{1}.6856$	$\overline{1}.6869$	$\overline{1}.6883$	$\overline{1}.6896$	$\overline{1}.6910$	$\overline{1}.6923$	$\overline{1}.6937$	$\overline{1}.6950$	$\overline{1}.6963$	$\overline{1}.6977$	2	4	7	9	11
30	$\overline{1}.6990$	$\overline{1}.7003$	$\overline{1}.7016$	$\overline{1}.7029$	$\overline{1}.7042$	$\overline{1}.7055$	$\overline{1}.7068$	$\overline{1}.7080$	$\overline{1}.7093$	$\overline{1}.7106$	2	4	6	9	11
31	$\overline{1}.7118$	$\overline{1}.7131$	$\overline{1}.7144$	$\overline{1}.7156$	$\overline{1}.7168$	$\overline{1}.7181$	$\overline{1}.7193$	$\overline{1}.7205$	$\overline{1}.7218$	$\overline{1}.7230$	2	4	6	8	10
32	$\overline{1}.7242$	$\overline{1}.7254$	$\overline{1}.7266$	$\overline{1}.7278$	$\overline{1}.7290$	$\overline{1}.7302$	$\overline{1}.7314$	$\overline{1}.7326$	$\overline{1}.7338$	$\overline{1}.7349$	2	4	6	8	10
33	$\overline{1}.7361$	$\overline{1}.7373$	$\overline{1}.7384$	$\overline{1}.7396$	$\overline{1}.7407$	$\overline{1}.7419$	$\overline{1}.7430$	$\overline{1}.7442$	$\overline{1}.7453$	$\overline{1}.7464$	2	4	6	8	10
34	$\overline{1}.7476$	$\overline{1}.7487$	$\overline{1}.7498$	$\overline{1}.7509$	$\overline{1}.7520$	$\overline{1}.7531$	$\overline{1}.7542$	$\overline{1}.7553$	$\overline{1}.7564$	$\overline{1}.7575$	2	4	6	7	9
35	$\overline{1}.7586$	$\overline{1}.7597$	$\overline{1}.7607$	$\overline{1}.7618$	$\overline{1}.7629$	$\overline{1}.7640$	$\overline{1}.7650$	$\overline{1}.7661$	$\overline{1}.7671$	$\overline{1}.7682$	2	4	5	7	9
36	$\overline{1}.7692$	$\overline{1}.7703$	$\overline{1}.7713$	$\overline{1}.7723$	$\overline{1}.7734$	$\overline{1}.7744$	$\overline{1}.7754$	$\overline{1}.7764$	$\overline{1}.7774$	$\overline{1}.7785$	2	3	5	7	9
37	$\overline{1}.7795$	$\overline{1}.7805$	$\overline{1}.7815$	$\overline{1}.7825$	$\overline{1}.7835$	$\overline{1}.7844$	$\overline{1}.7854$	$\overline{1}.7864$	$\overline{1}.7874$	$\overline{1}.7884$	2	3	5	7	8
38	$\overline{1}.7893$	$\overline{1}.7903$	$\overline{1}.7913$	$\overline{1}.7922$	$\overline{1}.7932$	$\overline{1}.7941$	$\overline{1}.7951$	$\overline{1}.7960$	$\overline{1}.7970$	$\overline{1}.7979$	2	3	5	6	8
39	$\overline{1}.7989$	$\overline{1}.7998$	$\overline{1}.8007$	$\overline{1}.8017$	$\overline{1}.8026$	$\overline{1}.8035$	$\overline{1}.8044$	$\overline{1}.8053$	$\overline{1}.8063$	$\overline{1}.8072$	2	3	5	6	8
40	$\overline{1}.8081$	$\overline{1}.8090$	$\overline{1}.8099$	$\overline{1}.8108$	$\overline{1}.8117$	$\overline{1}.8125$	$\overline{1}.8134$	$\overline{1}.8143$	$\overline{1}.8152$	$\overline{1}.8161$	1	3	4	6	7
41	$\overline{1}.8169$	$\overline{1}.8178$	$\overline{1}.8187$	$\overline{1}.8195$	$\overline{1}.8204$	$\overline{1}.8213$	$\overline{1}.8221$	$\overline{1}.8230$	$\overline{1}.8238$	$\overline{1}.8247$	1	3	4	6	7
42	$\overline{1}.8255$	$\overline{1}.8264$	$\overline{1}.8272$	$\overline{1}.8280$	$\overline{1}.8289$	$\overline{1}.8297$	$\overline{1}.8305$	$\overline{1}.8313$	$\overline{1}.8322$	$\overline{1}.8330$	1	3	4	6	7
43	$\overline{1}.8338$	$\overline{1}.8346$	$\overline{1}.8354$	$\overline{1}.8362$	$\overline{1}.8370$	$\overline{1}.8378$	$\overline{1}.8386$	$\overline{1}.8394$	$\overline{1}.8402$	$\overline{1}.8410$	1	3	4	5	7
44	$\overline{1}.8418$	$\overline{1}.8426$	$\overline{1}.8433$	$\overline{1}.8441$	$\overline{1}.8449$	$\overline{1}.8457$	$\overline{1}.8464$	$\overline{1}.8472$	$\overline{1}.8480$	$\overline{1}.8487$	1	3	4	5	6

To find 28.25 × sin 39° 17'

$$\log(28.25 \times \sin 39° 17') = \log 28.25 + \log \sin 39° 17'$$
$$= 1.4510 + \overline{1}.8015 = 1.2525$$

From the anti-log tables the answer is 17.88.

Logarithms of Sines

°	0' 0.0°	6' 0.1°	12' 0.2°	18' 0.3°	24' 0.4°	30' 0.5°	36' 0.6°	42' 0.7°	48' 0.8°	54' 0.9°	1'	2'	3'	4'	5'
45	$\bar{1}$.8495	$\bar{1}$.8502	$\bar{1}$.8510	$\bar{1}$.8517	$\bar{1}$.8525	$\bar{1}$.8532	$\bar{1}$.8540	$\bar{1}$.8547	$\bar{1}$.8555	$\bar{1}$.8562	1	2	4	5	6
46	$\bar{1}$.8569	$\bar{1}$.8577	$\bar{1}$.8584	$\bar{1}$.8591	$\bar{1}$.8598	$\bar{1}$.8606	$\bar{1}$.8613	$\bar{1}$.8620	$\bar{1}$.8627	$\bar{1}$.8634	1	2	4	5	6
47	$\bar{1}$.8641	$\bar{1}$.8648	$\bar{1}$.8655	$\bar{1}$.8662	$\bar{1}$.8669	$\bar{1}$.8676	$\bar{1}$.8683	$\bar{1}$.8690	$\bar{1}$.8697	$\bar{1}$.8704	1	2	3	5	6
48	$\bar{1}$.8711	$\bar{1}$.8718	$\bar{1}$.8724	$\bar{1}$.8731	$\bar{1}$.8738	$\bar{1}$.8745	$\bar{1}$.8751	$\bar{1}$.8758	$\bar{1}$.8765	$\bar{1}$.8771	1	2	3	4	6
49	$\bar{1}$.8778	$\bar{1}$.8784	$\bar{1}$.8791	$\bar{1}$.8797	$\bar{1}$.8804	$\bar{1}$.8810	$\bar{1}$.8817	$\bar{1}$.8823	$\bar{1}$.8830	$\bar{1}$.8836	1	2	3	4	5
50	$\bar{1}$.8843	$\bar{1}$.8849	$\bar{1}$.8855	$\bar{1}$.8862	$\bar{1}$.8868	$\bar{1}$.8874	$\bar{1}$.8880	$\bar{1}$.8887	$\bar{1}$.8893	$\bar{1}$.8899	1	2	3	4	5
51	$\bar{1}$.8905	$\bar{1}$.8911	$\bar{1}$.8917	$\bar{1}$.8923	$\bar{1}$.8929	$\bar{1}$.8935	$\bar{1}$.8941	$\bar{1}$.8947	$\bar{1}$.8953	$\bar{1}$.8959	1	2	3	4	5
52	$\bar{1}$.8965	$\bar{1}$.8971	$\bar{1}$.8977	$\bar{1}$.8983	$\bar{1}$.8989	$\bar{1}$.8995	$\bar{1}$.9000	$\bar{1}$.9006	$\bar{1}$.9012	$\bar{1}$.9018	1	2	3	4	5
53	$\bar{1}$.9023	$\bar{1}$.9029	$\bar{1}$.9035	$\bar{1}$.9041	$\bar{1}$.9046	$\bar{1}$.9052	$\bar{1}$.9057	$\bar{1}$.9063	$\bar{1}$.9069	$\bar{1}$.9074	1	2	3	4	5
54	$\bar{1}$.9080	$\bar{1}$.9085	$\bar{1}$.9091	$\bar{1}$.9096	$\bar{1}$.9101	$\bar{1}$.9107	$\bar{1}$.9112	$\bar{1}$.9118	$\bar{1}$.9123	$\bar{1}$.9128	1	2	3	4	5
55	$\bar{1}$.9134	$\bar{1}$.9139	$\bar{1}$.9144	$\bar{1}$.9149	$\bar{1}$.9155	$\bar{1}$.9160	$\bar{1}$.9165	$\bar{1}$.9170	$\bar{1}$.9175	$\bar{1}$.9181	1	2	3	3	4
56	$\bar{1}$.9186	$\bar{1}$.9191	$\bar{1}$.9196	$\bar{1}$.9201	$\bar{1}$.9206	$\bar{1}$.9211	$\bar{1}$.9216	$\bar{1}$.9221	$\bar{1}$.9226	$\bar{1}$.9231	1	2	3	3	4
57	$\bar{1}$.9236	$\bar{1}$.9241	$\bar{1}$.9246	$\bar{1}$.9251	$\bar{1}$.9255	$\bar{1}$.9260	$\bar{1}$.9265	$\bar{1}$.9270	$\bar{1}$.9275	$\bar{1}$.9279	1	2	2	3	4
58	$\bar{1}$.9284	$\bar{1}$.9289	$\bar{1}$.9294	$\bar{1}$.9298	$\bar{1}$.9303	$\bar{1}$.9308	$\bar{1}$.9312	$\bar{1}$.9317	$\bar{1}$.9322	$\bar{1}$.9326	1	2	2	3	4
59	$\bar{1}$.9331	$\bar{1}$.9335	$\bar{1}$.9340	$\bar{1}$.9344	$\bar{1}$.9349	$\bar{1}$.9353	$\bar{1}$.9358	$\bar{1}$.9362	$\bar{1}$.9367	$\bar{1}$.9371	1	1	2	3	4
60	$\bar{1}$.9375	$\bar{1}$.9380	$\bar{1}$.9384	$\bar{1}$.9388	$\bar{1}$.9393	$\bar{1}$.9397	$\bar{1}$.9401	$\bar{1}$.9406	$\bar{1}$.9410	$\bar{1}$.9414	1	1	2	3	4
61	$\bar{1}$.9418	$\bar{1}$.9422	$\bar{1}$.9427	$\bar{1}$.9431	$\bar{1}$.9435	$\bar{1}$.9439	$\bar{1}$.9443	$\bar{1}$.9447	$\bar{1}$.9451	$\bar{1}$.9455	1	1	2	3	3
62	$\bar{1}$.9459	$\bar{1}$.9463	$\bar{1}$.9467	$\bar{1}$.9471	$\bar{1}$.9475	$\bar{1}$.9479	$\bar{1}$.9483	$\bar{1}$.9487	$\bar{1}$.9491	$\bar{1}$.9495	1	1	2	3	3
63	$\bar{1}$.9499	$\bar{1}$.9503	$\bar{1}$.9506	$\bar{1}$.9510	$\bar{1}$.9514	$\bar{1}$.9518	$\bar{1}$.9522	$\bar{1}$.9525	$\bar{1}$.9529	$\bar{1}$.9533	1	1	2	3	3
64	$\bar{1}$.9537	$\bar{1}$.9540	$\bar{1}$.9544	$\bar{1}$.9548	$\bar{1}$.9551	$\bar{1}$.9555	$\bar{1}$.9558	$\bar{1}$.9562	$\bar{1}$.9566	$\bar{1}$.9569	1	1	2	2	3
65	$\bar{1}$.9573	$\bar{1}$.9576	$\bar{1}$.9580	$\bar{1}$.9583	$\bar{1}$.9587	$\bar{1}$.9590	$\bar{1}$.9594	$\bar{1}$.9597	$\bar{1}$.9601	$\bar{1}$.9604	1	1	2	2	3
66	$\bar{1}$.9607	$\bar{1}$.9611	$\bar{1}$.9614	$\bar{1}$.9617	$\bar{1}$.9621	$\bar{1}$.9624	$\bar{1}$.9627	$\bar{1}$.9631	$\bar{1}$.9634	$\bar{1}$.9637	1	1	2	2	3
67	$\bar{1}$.9640	$\bar{1}$.9643	$\bar{1}$.9647	$\bar{1}$.9650	$\bar{1}$.9653	$\bar{1}$.9656	$\bar{1}$.9659	$\bar{1}$.9662	$\bar{1}$.9666	$\bar{1}$.9669	1	1	2	2	3
68	$\bar{1}$.9672	$\bar{1}$.9675	$\bar{1}$.9678	$\bar{1}$.9681	$\bar{1}$.9684	$\bar{1}$.9687	$\bar{1}$.9690	$\bar{1}$.9693	$\bar{1}$.9696	$\bar{1}$.9699	1	1	2	2	2
69	$\bar{1}$.9702	$\bar{1}$.9704	$\bar{1}$.9707	$\bar{1}$.9710	$\bar{1}$.9713	$\bar{1}$.9716	$\bar{1}$.9719	$\bar{1}$.9722	$\bar{1}$.9724	$\bar{1}$.9727	0	1	1	2	2
70	$\bar{1}$.9730	$\bar{1}$.9733	$\bar{1}$.9735	$\bar{1}$.9738	$\bar{1}$.9741	$\bar{1}$.9743	$\bar{1}$.9746	$\bar{1}$.9749	$\bar{1}$.9751	$\bar{1}$.9754	0	1	1	2	2
71	$\bar{1}$.9757	$\bar{1}$.9759	$\bar{1}$.9762	$\bar{1}$.9764	$\bar{1}$.9767	$\bar{1}$.9770	$\bar{1}$.9772	$\bar{1}$.9775	$\bar{1}$.9777	$\bar{1}$.9780	0	1	1	2	2
72	$\bar{1}$.9782	$\bar{1}$.9785	$\bar{1}$.9787	$\bar{1}$.9789	$\bar{1}$.9792	$\bar{1}$.9794	$\bar{1}$.9797	$\bar{1}$.9799	$\bar{1}$.9801	$\bar{1}$.9804	0	1	1	2	2
73	$\bar{1}$.9806	$\bar{1}$.9808	$\bar{1}$.9811	$\bar{1}$.9813	$\bar{1}$.9815	$\bar{1}$.9817	$\bar{1}$.9820	$\bar{1}$.9822	$\bar{1}$.9824	$\bar{1}$.9826	0	1	1	1	2
74	$\bar{1}$.9828	$\bar{1}$.9831	$\bar{1}$.9833	$\bar{1}$.9835	$\bar{1}$.9837	$\bar{1}$.9839	$\bar{1}$.9841	$\bar{1}$.9843	$\bar{1}$.9845	$\bar{1}$.9847	0	1	1	1	2
75	$\bar{1}$.9849	$\bar{1}$.9851	$\bar{1}$.9853	$\bar{1}$.9855	$\bar{1}$.9857	$\bar{1}$.9859	$\bar{1}$.9861	$\bar{1}$.9863	$\bar{1}$.9865	$\bar{1}$.9867	0	1	1	1	2
76	$\bar{1}$.9869	$\bar{1}$.9871	$\bar{1}$.9873	$\bar{1}$.9875	$\bar{1}$.9876	$\bar{1}$.9878	$\bar{1}$.9880	$\bar{1}$.9882	$\bar{1}$.9884	$\bar{1}$.9885	0	1	1	1	2
77	$\bar{1}$.9887	$\bar{1}$.9889	$\bar{1}$.9891	$\bar{1}$.9892	$\bar{1}$.9894	$\bar{1}$.9896	$\bar{1}$.9897	$\bar{1}$.9899	$\bar{1}$.9901	$\bar{1}$.9902	0	1	1	1	1
78	$\bar{1}$.9904	$\bar{1}$.9906	$\bar{1}$.9907	$\bar{1}$.9909	$\bar{1}$.9910	$\bar{1}$.9912	$\bar{1}$.9913	$\bar{1}$.9915	$\bar{1}$.9916	$\bar{1}$.9918	0	1	1	1	1
79	$\bar{1}$.9919	$\bar{1}$.9921	$\bar{1}$.9922	$\bar{1}$.9924	$\bar{1}$.9925	$\bar{1}$.9927	$\bar{1}$.9928	$\bar{1}$.9929	$\bar{1}$.9931	$\bar{1}$.9932	0	0	1	1	1
80	$\bar{1}$.9934	$\bar{1}$.9935	$\bar{1}$.9936	$\bar{1}$.9937	$\bar{1}$.9939	$\bar{1}$.9940	$\bar{1}$.9941	$\bar{1}$.9943	$\bar{1}$.9944	$\bar{1}$.9945	0	0	1	1	1
81	$\bar{1}$.9946	$\bar{1}$.9947	$\bar{1}$.9949	$\bar{1}$.9950	$\bar{1}$.9951	$\bar{1}$.9952	$\bar{1}$.9953	$\bar{1}$.9954	$\bar{1}$.9955	$\bar{1}$.9956	0	0	1	1	1
82	$\bar{1}$.9958	$\bar{1}$.9959	$\bar{1}$.9960	$\bar{1}$.9961	$\bar{1}$.9962	$\bar{1}$.9963	$\bar{1}$.9964	$\bar{1}$.9965	$\bar{1}$.9966	$\bar{1}$.9967	0	0	0	1	1
83	$\bar{1}$.9968	$\bar{1}$.9968	$\bar{1}$.9969	$\bar{1}$.9970	$\bar{1}$.9971	$\bar{1}$.9972	$\bar{1}$.9973	$\bar{1}$.9974	$\bar{1}$.9975	$\bar{1}$.9975	0	0	0	1	1
84	$\bar{1}$.9976	$\bar{1}$.9977	$\bar{1}$.9978	$\bar{1}$.9978	$\bar{1}$.9979	$\bar{1}$.9980	$\bar{1}$.9981	$\bar{1}$.9981	$\bar{1}$.9982	$\bar{1}$.9983	0	0	0	0	1
85	$\bar{1}$.9983	$\bar{1}$.9984	$\bar{1}$.9985	$\bar{1}$.9985	$\bar{1}$.9986	$\bar{1}$.9987	$\bar{1}$.9987	$\bar{1}$.9988	$\bar{1}$.9988	$\bar{1}$.9989	0	0	0	0	0
86	$\bar{1}$.9989	$\bar{1}$.9990	$\bar{1}$.9990	$\bar{1}$.9991	$\bar{1}$.9991	$\bar{1}$.9992	$\bar{1}$.9992	$\bar{1}$.9993	$\bar{1}$.9993	$\bar{1}$.9994	0	0	0	0	0
87	$\bar{1}$.9994	$\bar{1}$.9994	$\bar{1}$.9995	$\bar{1}$.9995	$\bar{1}$.9996	$\bar{1}$.9996	$\bar{1}$.9996	$\bar{1}$.9996	$\bar{1}$.9997	$\bar{1}$.9997	0	0	0	0	0
88	$\bar{1}$.9997	$\bar{1}$.9998	$\bar{1}$.9998	$\bar{1}$.9998	$\bar{1}$.9998	$\bar{1}$.9999	$\bar{1}$.9999	$\bar{1}$.9999	$\bar{1}$.9999	$\bar{1}$.9999	0	0	0	0	0
89	$\bar{1}$.9999	$\bar{1}$.9999	0.0000	0.0000	0.0000	0.0000	0.0000	0.0000	0.0000	0.0000					
90	0.0000														

To find the angle A given that $\sin A = \dfrac{19.16}{23.45}$

$$\log \sin A = \log\left(\frac{19.16}{23.45}\right) = \log 19.16 - \log 23.45$$
$$= 1.2824 - 1.3701 = \bar{1}.9123$$

using the log sine tables

$$A = 54° 48'$$

Logarithms of Cosines

Numbers in difference columns to be *subtracted*, not added.

°	0' 0.0°	6' 0.1°	12' 0.2°	18' 0.3°	24' 0.4°	30' 0.5°	36' 0.6°	42' 0.7°	48' 0.8°	54' 0.9°	1'	2'	3'	4'	5'
0	0.0000	0.0000	0.0000	0.0000	0.0000	0.0000	0.0000	0.0000	0.0000	1̄.9999	0	0	0	0	0
1	1̄.9999	1̄.9999	1̄.9999	1̄.9999	1̄.9999	1̄.9999	1̄.9998	1̄.9998	1̄.9998	1̄.9998	0	0	0	0	0
2	1̄.9997	1̄.9997	1̄.9997	1̄.9996	1̄.9996	1̄.9996	1̄.9996	1̄.9995	1̄.9995	1̄.9994	0	0	0	0	0
3	1̄.9994	1̄.9994	1̄.9993	1̄.9993	1̄.9992	1̄.9992	1̄.9991	1̄.9991	1̄.9990	1̄.9990	0	0	0	0	0
4	1̄.9989	1̄.9989	1̄.9988	1̄.9988	1̄.9987	1̄.9987	1̄.9986	1̄.9985	1̄.9985	1̄.9984	0	0	0	0	0
5	1̄.9983	1̄.9983	1̄.9982	1̄.9981	1̄.9981	1̄.9980	1̄.9979	1̄.9978	1̄.9978	1̄.9977	0	0	0	0	1
6	1̄.9976	1̄.9975	1̄.9975	1̄.9974	1̄.9973	1̄.9972	1̄.9971	1̄.9970	1̄.9969	1̄.9968	0	0	0	1	1
7	1̄.9968	1̄.9967	1̄.9966	1̄.9965	1̄.9964	1̄.9963	1̄.9962	1̄.9961	1̄.9960	1̄.9959	0	0	0	1	1
8	1̄.9958	1̄.9956	1̄.9955	1̄.9954	1̄.9953	1̄.9952	1̄.9951	1̄.9950	1̄.9949	1̄.9947	0	0	1	1	1
9	1̄.9946	1̄.9945	1̄.9944	1̄.9943	1̄.9941	1̄.9940	1̄.9939	1̄.9937	1̄.9936	1̄.9935	0	0	1	1	1
10	1̄.9934	1̄.9932	1̄.9931	1̄.9929	1̄.9928	1̄.9927	1̄.9925	1̄.9924	1̄.9922	1̄.9921	0	0	1	1	1
11	1̄.9919	1̄.9918	1̄.9916	1̄.9915	1̄.9913	1̄.9912	1̄.9910	1̄.9909	1̄.9907	1̄.9906	0	1	1	1	1
12	1̄.9904	1̄.9902	1̄.9901	1̄.9899	1̄.9897	1̄.9896	1̄.9894	1̄.9892	1̄.9891	1̄.9889	0	1	1	1	1
13	1̄.9887	1̄.9885	1̄.9884	1̄.9882	1̄.9880	1̄.9878	1̄.9876	1̄.9875	1̄.9873	1̄.9871	0	1	1	1	2
14	1̄.9869	1̄.9867	1̄.9865	1̄.9863	1̄.9861	1̄.9859	1̄.9857	1̄.9855	1̄.9853	1̄.9851	0	1	1	1	2
15	1̄.9849	1̄.9847	1̄.9845	1̄.9843	1̄.9841	1̄.9839	1̄.9837	1̄.9835	1̄.9833	1̄.9831	0	1	1	1	2
16	1̄.9828	1̄.9826	1̄.9824	1̄.9822	1̄.9820	1̄.9817	1̄.9815	1̄.9813	1̄.9811	1̄.9808	0	1	1	1	2
17	1̄.9806	1̄.9804	1̄.9801	1̄.9799	1̄.9797	1̄.9794	1̄.9792	1̄.9789	1̄.9787	1̄.9785	0	1	1	2	2
18	1̄.9782	1̄.9780	1̄.9777	1̄.9775	1̄.9772	1̄.9770	1̄.9767	1̄.9764	1̄.9762	1̄.9759	0	1	1	2	2
19	1̄.9757	1̄.9754	1̄.9751	1̄.9749	1̄.9746	1̄.9743	1̄.9741	1̄.9738	1̄.9735	1̄.9733	0	1	1	2	2
20	1̄.9730	1̄.9727	1̄.9724	1̄.9722	1̄.9719	1̄.9716	1̄.9713	1̄.9710	1̄.9707	1̄.9704	0	1	1	2	2
21	1̄.9702	1̄.9699	1̄.9696	1̄.9693	1̄.9690	1̄.9687	1̄.9684	1̄.9681	1̄.9678	1̄.9675	1	1	2	2	2
22	1̄.9672	1̄.9669	1̄.9666	1̄.9662	1̄.9659	1̄.9656	1̄.9653	1̄.9650	1̄.9647	1̄.9643	1	1	2	2	3
23	1̄.9640	1̄.9637	1̄.9634	1̄.9631	1̄.9627	1̄.9624	1̄.9621	1̄.9617	1̄.9614	1̄.9611	1	1	2	2	3
24	1̄.9607	1̄.9604	1̄.9601	1̄.9597	1̄.9594	1̄.9590	1̄.9587	1̄.9583	1̄.9580	1̄.9576	1	1	2	2	3
25	1̄.9573	1̄.9569	1̄.9566	1̄.9562	1̄.9558	1̄.9555	1̄.9551	1̄.9548	1̄.9544	1̄.9540	1	1	2	2	3
26	1̄.9537	1̄.9533	1̄.9529	1̄.9525	1̄.9522	1̄.9518	1̄.9514	1̄.9510	1̄.9506	1̄.9503	1	1	2	3	3
27	1̄.9499	1̄.9495	1̄.9491	1̄.9487	1̄.9483	1̄.9479	1̄.9475	1̄.9471	1̄.9467	1̄.9463	1	1	2	3	3
28	1̄.9459	1̄.9455	1̄.9451	1̄.9447	1̄.9443	1̄.9439	1̄.9435	1̄.9431	1̄.9427	1̄.9422	1	1	2	3	3
29	1̄.9418	1̄.9414	1̄.9410	1̄.9406	1̄.9401	1̄.9397	1̄.9393	1̄.9388	1̄.9384	1̄.9380	1	1	2	3	4
30	1̄.9375	1̄.9371	1̄.9367	1̄.9362	1̄.9358	1̄.9353	1̄.9349	1̄.9344	1̄.9340	1̄.9335	1	1	2	3	4
31	1̄.9331	1̄.9326	1̄.9322	1̄.9317	1̄.9312	1̄.9308	1̄.9303	1̄.9298	1̄.9294	1̄.9289	1	2	2	3	4
32	1̄.9284	1̄.9279	1̄.9275	1̄.9270	1̄.9265	1̄.9260	1̄.9255	1̄.9251	1̄.9246	1̄.9241	1	2	2	3	4
33	1̄.9236	1̄.9231	1̄.9226	1̄.9221	1̄.9216	1̄.9211	1̄.9206	1̄.9201	1̄.9196	1̄.9191	1	2	3	3	4
34	1̄.9186	1̄.9181	1̄.9175	1̄.9170	1̄.9165	1̄.9160	1̄.9155	1̄.9149	1̄.9144	1̄.9139	1	2	3	3	4
35	1̄.9134	1̄.9128	1̄.9123	1̄.9118	1̄.9112	1̄.9107	1̄.9101	1̄.9096	1̄.9091	1̄.9085	1	2	3	4	5
36	1̄.9080	1̄.9074	1̄.9069	1̄.9063	1̄.9057	1̄.9052	1̄.9046	1̄.9041	1̄.9035	1̄.9029	1	2	3	4	5
37	1̄.9023	1̄.9018	1̄.9012	1̄.9006	1̄.9000	1̄.8995	1̄.8989	1̄.8983	1̄.8977	1̄.8971	1	2	3	4	5
38	1̄.8965	1̄.8959	1̄.8953	1̄.8947	1̄.8941	1̄.8935	1̄.8929	1̄.8923	1̄.8917	1̄.8911	1	2	3	4	5
39	1̄.8905	1̄.8899	1̄.8893	1̄.8887	1̄.8880	1̄.8874	1̄.8868	1̄.8862	1̄.8855	1̄.8849	1	2	3	4	5
40	1̄.8843	1̄.8836	1̄.8830	1̄.8823	1̄.8817	1̄.8810	1̄.8804	1̄.8797	1̄.8791	1̄.8784	1	2	3	4	5
41	1̄.8778	1̄.8771	1̄.8765	1̄.8758	1̄.8751	1̄.8745	1̄.8738	1̄.8731	1̄.8724	1̄.8718	1	2	3	4	6
42	1̄.8711	1̄.8704	1̄.8697	1̄.8690	1̄.8683	1̄.8676	1̄.8669	1̄.8662	1̄.8655	1̄.8648	1	2	4	5	6
43	1̄.8641	1̄.8634	1̄.8627	1̄.8620	1̄.8613	1̄.8606	1̄.8598	1̄.8591	1̄.8584	1̄.8577	1	2	4	5	6
44	1̄.8569	1̄.8562	1̄.8555	1̄.8547	1̄.8540	1̄.8532	1̄.8525	1̄.8517	1̄.8510	1̄.8502	1	2	4	5	6

To find $23.05 \times \cos 16° 51'$.

$$\log(23.05 \times \cos 16° 51') = \log 23.05 \times \log \cos 16° 51'$$
$$= 1.3626 \times \bar{1}.9810$$
$$= 1.3436$$

From the anti-log tables the answer is 22.06.

Logarithms of Cosines

Numbers in difference columns to be *subtracted*, not added.

°	0′ 0.0°	6′ 0.1°	12′ 0.2°	18′ 0.3°	24′ 0.4°	30′ 0.5°	36′ 0.6°	42′ 0.7°	48′ 0.8°	54′ 0.9°	1′	2′	3′	4′	5′
45	$\bar{1}$.8495	$\bar{1}$.8487	$\bar{1}$.8480	$\bar{1}$.8472	$\bar{1}$.8464	$\bar{1}$.8457	$\bar{1}$.8449	$\bar{1}$.8441	$\bar{1}$.8433	$\bar{1}$.8426	1	3	4	5	6
46	$\bar{1}$.8418	$\bar{1}$.8410	$\bar{1}$.8402	$\bar{1}$.8394	$\bar{1}$.8386	$\bar{1}$.8378	$\bar{1}$.8370	$\bar{1}$.8362	$\bar{1}$.8354	$\bar{1}$.8346	1	3	4	5	7
47	$\bar{1}$.8338	$\bar{1}$.8330	$\bar{1}$.8322	$\bar{1}$.8313	$\bar{1}$.8305	$\bar{1}$.8297	$\bar{1}$.8289	$\bar{1}$.8280	$\bar{1}$.8272	$\bar{1}$.8264	1	3	4	6	7
48	$\bar{1}$.8255	$\bar{1}$.8247	$\bar{1}$.8238	$\bar{1}$.8230	$\bar{1}$.8221	$\bar{1}$.8213	$\bar{1}$.8204	$\bar{1}$.8195	$\bar{1}$.8187	$\bar{1}$.8178	1	3	4	6	7
49	$\bar{1}$.8169	$\bar{1}$.8161	$\bar{1}$.8152	$\bar{1}$.8143	$\bar{1}$.8134	$\bar{1}$.8125	$\bar{1}$.8117	$\bar{1}$.8108	$\bar{1}$.8099	$\bar{1}$.8090	1	3	4	6	7
50	$\bar{1}$.8081	$\bar{1}$.8072	$\bar{1}$.8063	$\bar{1}$.8053	$\bar{1}$.8044	$\bar{1}$.8035	$\bar{1}$.8026	$\bar{1}$.8017	$\bar{1}$.8007	$\bar{1}$.7998	2	3	5	6	8
51	$\bar{1}$.7989	$\bar{1}$.7979	$\bar{1}$.7970	$\bar{1}$.7960	$\bar{1}$.7951	$\bar{1}$.7941	$\bar{1}$.7932	$\bar{1}$.7922	$\bar{1}$.7913	$\bar{1}$.7903	2	3	5	6	8
52	$\bar{1}$.7893	$\bar{1}$.7884	$\bar{1}$.7874	$\bar{1}$.7864	$\bar{1}$.7854	$\bar{1}$.7844	$\bar{1}$.7835	$\bar{1}$.7825	$\bar{1}$.7815	$\bar{1}$.7805	2	3	5	7	8
53	$\bar{1}$.7795	$\bar{1}$.7785	$\bar{1}$.7774	$\bar{1}$.7764	$\bar{1}$.7754	$\bar{1}$.7744	$\bar{1}$.7734	$\bar{1}$.7723	$\bar{1}$.7713	$\bar{1}$.7703	2	3	5	7	9
54	$\bar{1}$.7692	$\bar{1}$.7682	$\bar{1}$.7671	$\bar{1}$.7661	$\bar{1}$.7650	$\bar{1}$.7640	$\bar{1}$.7629	$\bar{1}$.7618	$\bar{1}$.7607	$\bar{1}$.7597	2	4	5	7	9
55	$\bar{1}$.7586	$\bar{1}$.7575	$\bar{1}$.7564	$\bar{1}$.7553	$\bar{1}$.7542	$\bar{1}$.7531	$\bar{1}$.7520	$\bar{1}$.7509	$\bar{1}$.7498	$\bar{1}$.7487	2	4	6	7	9
56	$\bar{1}$.7476	$\bar{1}$.7464	$\bar{1}$.7453	$\bar{1}$.7442	$\bar{1}$.7430	$\bar{1}$.7419	$\bar{1}$.7407	$\bar{1}$.7396	$\bar{1}$.7384	$\bar{1}$.7373	2	4	6	8	10
57	$\bar{1}$.7361	$\bar{1}$.7349	$\bar{1}$.7338	$\bar{1}$.7326	$\bar{1}$.7314	$\bar{1}$.7302	$\bar{1}$.7290	$\bar{1}$.7278	$\bar{1}$.7266	$\bar{1}$.7254	2	4	6	8	10
58	$\bar{1}$.7242	$\bar{1}$.7230	$\bar{1}$.7218	$\bar{1}$.7205	$\bar{1}$.7193	$\bar{1}$.7181	$\bar{1}$.7168	$\bar{1}$.7156	$\bar{1}$.7144	$\bar{1}$.7131	2	4	6	8	10
59	$\bar{1}$.7118	$\bar{1}$.7106	$\bar{1}$.7093	$\bar{1}$.7080	$\bar{1}$.7068	$\bar{1}$.7055	$\bar{1}$.7042	$\bar{1}$.7029	$\bar{1}$.7016	$\bar{1}$.7003	2	4	6	9	11
60	$\bar{1}$.6990	$\bar{1}$.6977	$\bar{1}$.6963	$\bar{1}$.6950	$\bar{1}$.6937	$\bar{1}$.6923	$\bar{1}$.6910	$\bar{1}$.6896	$\bar{1}$.6883	$\bar{1}$.6869	2	4	7	9	11
61	$\bar{1}$.6856	$\bar{1}$.6842	$\bar{1}$.6828	$\bar{1}$.6814	$\bar{1}$.6801	$\bar{1}$.6787	$\bar{1}$.6773	$\bar{1}$.6759	$\bar{1}$.6744	$\bar{1}$.6730	2	5	7	9	12
62	$\bar{1}$.6716	$\bar{1}$.6702	$\bar{1}$.6687	$\bar{1}$.6673	$\bar{1}$.6659	$\bar{1}$.6644	$\bar{1}$.6629	$\bar{1}$.6615	$\bar{1}$.6600	$\bar{1}$.6585	2	5	7	10	12
63	$\bar{1}$.6570	$\bar{1}$.6556	$\bar{1}$.6541	$\bar{1}$.6526	$\bar{1}$.6510	$\bar{1}$.6495	$\bar{1}$.6480	$\bar{1}$.6465	$\bar{1}$.6449	$\bar{1}$.6434	3	5	8	10	13
64	$\bar{1}$.6418	$\bar{1}$.6403	$\bar{1}$.6387	$\bar{1}$.6371	$\bar{1}$.6356	$\bar{1}$.6340	$\bar{1}$.6324	$\bar{1}$.6308	$\bar{1}$.6292	$\bar{1}$.6276	3	5	8	11	13
65	$\bar{1}$.6259	$\bar{1}$.6243	$\bar{1}$.6227	$\bar{1}$.6210	$\bar{1}$.6194	$\bar{1}$.6177	$\bar{1}$.6161	$\bar{1}$.6144	$\bar{1}$.6127	$\bar{1}$.6110	3	6	8	11	14
66	$\bar{1}$.6093	$\bar{1}$.6076	$\bar{1}$.6059	$\bar{1}$.6042	$\bar{1}$.6024	$\bar{1}$.6007	$\bar{1}$.5990	$\bar{1}$.5972	$\bar{1}$.5954	$\bar{1}$.5937	3	6	9	12	14
67	$\bar{1}$.5919	$\bar{1}$.5901	$\bar{1}$.5883	$\bar{1}$.5865	$\bar{1}$.5847	$\bar{1}$.5828	$\bar{1}$.5810	$\bar{1}$.5792	$\bar{1}$.5773	$\bar{1}$.5754	3	6	9	12	15
68	$\bar{1}$.5736	$\bar{1}$.5717	$\bar{1}$.5698	$\bar{1}$.5679	$\bar{1}$.5660	$\bar{1}$.5641	$\bar{1}$.5621	$\bar{1}$.5602	$\bar{1}$.5583	$\bar{1}$.5563	3	6	10	13	16
69	$\bar{1}$.5543	$\bar{1}$.5523	$\bar{1}$.5504	$\bar{1}$.5484	$\bar{1}$.5463	$\bar{1}$.5443	$\bar{1}$.5423	$\bar{1}$.5402	$\bar{1}$.5382	$\bar{1}$.5361	3	7	10	13	17
70	$\bar{1}$.5341	$\bar{1}$.5320	$\bar{1}$.5299	$\bar{1}$.5278	$\bar{1}$.5256	$\bar{1}$.5235	$\bar{1}$.5213	$\bar{1}$.5192	$\bar{1}$.5170	$\bar{1}$.5148	4	7	11	14	18
71	$\bar{1}$.5126	$\bar{1}$.5104	$\bar{1}$.5082	$\bar{1}$.5060	$\bar{1}$.5037	$\bar{1}$.5015	$\bar{1}$.4992	$\bar{1}$.4969	$\bar{1}$.4946	$\bar{1}$.4923	4	8	11	15	19
72	$\bar{1}$.4900	$\bar{1}$.4876	$\bar{1}$.4853	$\bar{1}$.4829	$\bar{1}$.4805	$\bar{1}$.4781	$\bar{1}$.4757	$\bar{1}$.4733	$\bar{1}$.4709	$\bar{1}$.4684	4	8	12	16	20
73	$\bar{1}$.4659	$\bar{1}$.4634	$\bar{1}$.4609	$\bar{1}$.4584	$\bar{1}$.4559	$\bar{1}$.4533	$\bar{1}$.4508	$\bar{1}$.4482	$\bar{1}$.4456	$\bar{1}$.4430	4	9	13	17	21
74	$\bar{1}$.4403	$\bar{1}$.4377	$\bar{1}$.4350	$\bar{1}$.4323	$\bar{1}$.4296	$\bar{1}$.4269	$\bar{1}$.4242	$\bar{1}$.4214	$\bar{1}$.4186	$\bar{1}$.4158	5	9	14	18	23
75	$\bar{1}$.4130	$\bar{1}$.4102	$\bar{1}$.4073	$\bar{1}$.4044	$\bar{1}$.4015	$\bar{1}$.3986	$\bar{1}$.3957	$\bar{1}$.3927	$\bar{1}$.3897	$\bar{1}$.3867	5	10	15	20	24
76	$\bar{1}$.3837	$\bar{1}$.3806	$\bar{1}$.3775	$\bar{1}$.3745	$\bar{1}$.3713	$\bar{1}$.3682	$\bar{1}$.3650	$\bar{1}$.3618	$\bar{1}$.3586	$\bar{1}$.3554	5	11	16	21	26
77	$\bar{1}$.3521	$\bar{1}$.3488	$\bar{1}$.3455	$\bar{1}$.3421	$\bar{1}$.3387	$\bar{1}$.3353	$\bar{1}$.3319	$\bar{1}$.3284	$\bar{1}$.3250	$\bar{1}$.3214	6	11	17	23	28
78	$\bar{1}$.3179	$\bar{1}$.3143	$\bar{1}$.3107	$\bar{1}$.3070	$\bar{1}$.3034	$\bar{1}$.2997	$\bar{1}$.2959	$\bar{1}$.2921	$\bar{1}$.2883	$\bar{1}$.2845	6	12	19	25	31
79	$\bar{1}$.2806	$\bar{1}$.2767	$\bar{1}$.2727	$\bar{1}$.2687	$\bar{1}$.2647	$\bar{1}$.2606	$\bar{1}$.2565	$\bar{1}$.2524	$\bar{1}$.2482	$\bar{1}$.2439	7	14	20	27	34
80	$\bar{1}$.2397	$\bar{1}$.2353	$\bar{1}$.2310	$\bar{1}$.2266	$\bar{1}$.2221	$\bar{1}$.2176	$\bar{1}$.2131	$\bar{1}$.2085	$\bar{1}$.2038	$\bar{1}$.1991	8	15	23	30	38
81	$\bar{1}$.1943	$\bar{1}$.1895	$\bar{1}$.1847	$\bar{1}$.1797	$\bar{1}$.1747	$\bar{1}$.1697	$\bar{1}$.1646	$\bar{1}$.1594	$\bar{1}$.1542	$\bar{1}$.1489	8	17	25	34	42
82	$\bar{1}$.1436	$\bar{1}$.1381	$\bar{1}$.1326	$\bar{1}$.1271	$\bar{1}$.1214	$\bar{1}$.1157	$\bar{1}$.1099	$\bar{1}$.1040	$\bar{1}$.0981	$\bar{1}$.0920	10	19	29	38	48
83	$\bar{1}$.0859	$\bar{1}$.0797	$\bar{1}$.0734	$\bar{1}$.0670	$\bar{1}$.0605	$\bar{1}$.0539	$\bar{1}$.0472	$\bar{1}$.0403	$\bar{1}$.0334	$\bar{1}$.0264	11	22	33	44	55
84	$\bar{1}$.0192	$\bar{1}$.0120	$\bar{1}$.0046	$\bar{2}$.9970	$\bar{2}$.9894	$\bar{2}$.9816	$\bar{2}$.9736	$\bar{2}$.9655	$\bar{2}$.9573	$\bar{2}$.9489	13	26	39	52	65
85	$\bar{2}$.9403	$\bar{2}$.9315	$\bar{2}$.9226	$\bar{2}$.9135	$\bar{2}$.9042	$\bar{2}$.8946	$\bar{2}$.8849	$\bar{2}$.8749	$\bar{2}$.8647	$\bar{2}$.8543	16	32	48	64	80
86	$\bar{2}$.8436	$\bar{2}$.8326	$\bar{2}$.8213	$\bar{2}$.8098	$\bar{2}$.7979	$\bar{2}$.7857	$\bar{2}$.7731	$\bar{2}$.7602	$\bar{2}$.7468	$\bar{2}$.7330					
87	$\bar{2}$.7188	$\bar{2}$.7041	$\bar{2}$.6889	$\bar{2}$.6731	$\bar{2}$.6567	$\bar{2}$.6397	$\bar{2}$.6220	$\bar{2}$.6035	$\bar{2}$.5842	$\bar{2}$.5640					
88	$\bar{2}$.5428	$\bar{2}$.5206	$\bar{2}$.4971	$\bar{2}$.4723	$\bar{2}$.4459	$\bar{2}$.4179	$\bar{2}$.3880	$\bar{2}$.3558	$\bar{2}$.3210	$\bar{2}$.2832		Differences untrustworthy here			
89	$\bar{2}$.2419	$\bar{2}$.1961	$\bar{2}$.1450	$\bar{2}$.0870	$\bar{2}$.0200	$\bar{3}$.9408	$\bar{3}$.8439	$\bar{3}$.7190	$\bar{3}$.5429	$\bar{3}$.2419					
90	−∞														

To find the angle A given that $\cos A = \dfrac{20.23}{29.86}$

$$\log \cos A = \log\left(\frac{20.23}{29.86}\right) = \log 20.23 - \log 29.86$$
$$= 1.3060 - 1.4751 = \bar{1}.8309$$

Using the log cosine tables

$$A = 47° \, 21′$$

51

Logarithms of Tangents

°	0' 0.0°	6' 0.1°	12' 0.2°	18' 0.3°	24' 0.4°	30' 0.5°	36' 0.6°	42' 0.7°	48' 0.8°	54' 0.9°	1'	2'	3'	4'	5'
0	−∞	$\bar{3}$.2419	$\bar{3}$.5429	$\bar{3}$.7190	$\bar{3}$.8439	$\bar{3}$.9409	$\bar{2}$.0200	$\bar{2}$.0870	$\bar{2}$.1450	$\bar{2}$.1962		Differences			
1	$\bar{2}$.2419	$\bar{2}$.2833	$\bar{2}$.3211	$\bar{2}$.3559	$\bar{2}$.3881	$\bar{2}$.4181	$\bar{2}$.4461	$\bar{2}$.4725	$\bar{2}$.4973	$\bar{2}$.5208		untrustworthy			
2	$\bar{2}$.5431	$\bar{2}$.5643	$\bar{2}$.5845	$\bar{2}$.6038	$\bar{2}$.6223	$\bar{2}$.6401	$\bar{2}$.6571	$\bar{2}$.6736	$\bar{2}$.6894	$\bar{2}$.7046		here			
3	$\bar{2}$.7194	$\bar{2}$.7337	$\bar{2}$.7475	$\bar{2}$.7609	$\bar{2}$.7739	$\bar{2}$.7865	$\bar{2}$.7988	$\bar{2}$.8107	$\bar{2}$.8223	$\bar{2}$.8336					
4	$\bar{2}$.8446	$\bar{2}$.8554	$\bar{2}$.8659	$\bar{2}$.8762	$\bar{2}$.8862	$\bar{2}$.8960	$\bar{2}$.9056	$\bar{2}$.9150	$\bar{2}$.9241	$\bar{2}$.9331	16	32	48	64	81
5	$\bar{2}$.9420	$\bar{2}$.9506	$\bar{2}$.9591	$\bar{2}$.9674	$\bar{2}$.9756	$\bar{2}$.9836	$\bar{2}$.9915	$\bar{2}$.9992	$\bar{1}$.0068	$\bar{1}$.0143	13	26	40	53	66
6	$\bar{1}$.0216	$\bar{1}$.0289	$\bar{1}$.0360	$\bar{1}$.0430	$\bar{1}$.0499	$\bar{1}$.0567	$\bar{1}$.0633	$\bar{1}$.0699	$\bar{1}$.0764	$\bar{1}$.0828	11	22	34	45	56
7	$\bar{1}$.0891	$\bar{1}$.0954	$\bar{1}$.1015	$\bar{1}$.1076	$\bar{1}$.1135	$\bar{1}$.1194	$\bar{1}$.1252	$\bar{1}$.1310	$\bar{1}$.1367	$\bar{1}$.1423	10	20	29	39	49
8	$\bar{1}$.1478	$\bar{1}$.1533	$\bar{1}$.1587	$\bar{1}$.1640	$\bar{1}$.1693	$\bar{1}$.1745	$\bar{1}$.1797	$\bar{1}$.1848	$\bar{1}$.1898	$\bar{1}$.1948	9	17	26	35	43
9	$\bar{1}$.1997	$\bar{1}$.2046	$\bar{1}$.2094	$\bar{1}$.2142	$\bar{1}$.2189	$\bar{1}$.2236	$\bar{1}$.2282	$\bar{1}$.2328	$\bar{1}$.2374	$\bar{1}$.2419	8	16	23	31	39
10	$\bar{1}$.2463	$\bar{1}$.2507	$\bar{1}$.2551	$\bar{1}$.2594	$\bar{1}$.2637	$\bar{1}$.2680	$\bar{1}$.2722	$\bar{1}$.2764	$\bar{1}$.2805	$\bar{1}$.2846	7	14	21	28	35
11	$\bar{1}$.2887	$\bar{1}$.2927	$\bar{1}$.2967	$\bar{1}$.3006	$\bar{1}$.3046	$\bar{1}$.3085	$\bar{1}$.3123	$\bar{1}$.3162	$\bar{1}$.3200	$\bar{1}$.3237	6	13	19	26	32
12	$\bar{1}$.3275	$\bar{1}$.3312	$\bar{1}$.3349	$\bar{1}$.3385	$\bar{1}$.3422	$\bar{1}$.3458	$\bar{1}$.3493	$\bar{1}$.3529	$\bar{1}$.3564	$\bar{1}$.3599	6	12	18	24	30
13	$\bar{1}$.3634	$\bar{1}$.3668	$\bar{1}$.3702	$\bar{1}$.3736	$\bar{1}$.3770	$\bar{1}$.3804	$\bar{1}$.3837	$\bar{1}$.3870	$\bar{1}$.3903	$\bar{1}$.3935	6	11	17	22	28
14	$\bar{1}$.3968	$\bar{1}$.4000	$\bar{1}$.4032	$\bar{1}$.4064	$\bar{1}$.4095	$\bar{1}$.4127	$\bar{1}$.4158	$\bar{1}$.4189	$\bar{1}$.4220	$\bar{1}$.4250	5	10	16	21	26
15	$\bar{1}$.4281	$\bar{1}$.4311	$\bar{1}$.4341	$\bar{1}$.4371	$\bar{1}$.4400	$\bar{1}$.4430	$\bar{1}$.4459	$\bar{1}$.4488	$\bar{1}$.4517	$\bar{1}$.4546	5	10	15	20	24
16	$\bar{1}$.4575	$\bar{1}$.4603	$\bar{1}$.4632	$\bar{1}$.4660	$\bar{1}$.4688	$\bar{1}$.4716	$\bar{1}$.4744	$\bar{1}$.4771	$\bar{1}$.4799	$\bar{1}$.4826	5	9	14	19	23
17	$\bar{1}$.4853	$\bar{1}$.4880	$\bar{1}$.4907	$\bar{1}$.4934	$\bar{1}$.4961	$\bar{1}$.4987	$\bar{1}$.5014	$\bar{1}$.5040	$\bar{1}$.5066	$\bar{1}$.5092	4	9	13	18	22
18	$\bar{1}$.5118	$\bar{1}$.5143	$\bar{1}$.5169	$\bar{1}$.5195	$\bar{1}$.5220	$\bar{1}$.5245	$\bar{1}$.5270	$\bar{1}$.5295	$\bar{1}$.5320	$\bar{1}$.5345	4	8	13	17	21
19	$\bar{1}$.5370	$\bar{1}$.5394	$\bar{1}$.5419	$\bar{1}$.5443	$\bar{1}$.5467	$\bar{1}$.5491	$\bar{1}$.5516	$\bar{1}$.5539	$\bar{1}$.5563	$\bar{1}$.5587	4	8	12	16	20
20	$\bar{1}$.5611	$\bar{1}$.5634	$\bar{1}$.5658	$\bar{1}$.5681	$\bar{1}$.5704	$\bar{1}$.5727	$\bar{1}$.5750	$\bar{1}$.5773	$\bar{1}$.5796	$\bar{1}$.5819	4	8	12	15	19
21	$\bar{1}$.5842	$\bar{1}$.5864	$\bar{1}$.5887	$\bar{1}$.5909	$\bar{1}$.5932	$\bar{1}$.5954	$\bar{1}$.5976	$\bar{1}$.5998	$\bar{1}$.6020	$\bar{1}$.6042	4	7	11	15	18
22	$\bar{1}$.6064	$\bar{1}$.6086	$\bar{1}$.6108	$\bar{1}$.6129	$\bar{1}$.6151	$\bar{1}$.6172	$\bar{1}$.6194	$\bar{1}$.6215	$\bar{1}$.6236	$\bar{1}$.6257	4	7	11	14	18
23	$\bar{1}$.6279	$\bar{1}$.6300	$\bar{1}$.6321	$\bar{1}$.6341	$\bar{1}$.6362	$\bar{1}$.6383	$\bar{1}$.6404	$\bar{1}$.6424	$\bar{1}$.6445	$\bar{1}$.6465	3	7	10	14	17
24	$\bar{1}$.6486	$\bar{1}$.6506	$\bar{1}$.6527	$\bar{1}$.6547	$\bar{1}$.6567	$\bar{1}$.6587	$\bar{1}$.6607	$\bar{1}$.6627	$\bar{1}$.6647	$\bar{1}$.6667	3	7	10	13	17
25	$\bar{1}$.6687	$\bar{1}$.6706	$\bar{1}$.6726	$\bar{1}$.6746	$\bar{1}$.6765	$\bar{1}$.6785	$\bar{1}$.6804	$\bar{1}$.6824	$\bar{1}$.6843	$\bar{1}$.6863	3	6	10	13	16
26	$\bar{1}$.6882	$\bar{1}$.6901	$\bar{1}$.6920	$\bar{1}$.6939	$\bar{1}$.6958	$\bar{1}$.6977	$\bar{1}$.6996	$\bar{1}$.7015	$\bar{1}$.7034	$\bar{1}$.7053	3	6	10	13	16
27	$\bar{1}$.7072	$\bar{1}$.7090	$\bar{1}$.7109	$\bar{1}$.7128	$\bar{1}$.7146	$\bar{1}$.7165	$\bar{1}$.7183	$\bar{1}$.7202	$\bar{1}$.7220	$\bar{1}$.7238	3	6	9	12	15
28	$\bar{1}$.7257	$\bar{1}$.7275	$\bar{1}$.7293	$\bar{1}$.7311	$\bar{1}$.7330	$\bar{1}$.7348	$\bar{1}$.7366	$\bar{1}$.7384	$\bar{1}$.7402	$\bar{1}$.7420	3	6	9	12	15
29	$\bar{1}$.7438	$\bar{1}$.7455	$\bar{1}$.7473	$\bar{1}$.7491	$\bar{1}$.7509	$\bar{1}$.7526	$\bar{1}$.7544	$\bar{1}$.7562	$\bar{1}$.7579	$\bar{1}$.7597	3	6	9	12	15
30	$\bar{1}$.7614	$\bar{1}$.7632	$\bar{1}$.7649	$\bar{1}$.7667	$\bar{1}$.7684	$\bar{1}$.7701	$\bar{1}$.7719	$\bar{1}$.7736	$\bar{1}$.7753	$\bar{1}$.7771	3	6	9	12	15
31	$\bar{1}$.7788	$\bar{1}$.7805	$\bar{1}$.7822	$\bar{1}$.7839	$\bar{1}$.7856	$\bar{1}$.7873	$\bar{1}$.7890	$\bar{1}$.7907	$\bar{1}$.7924	$\bar{1}$.7941	3	6	9	11	14
32	$\bar{1}$.7958	$\bar{1}$.7975	$\bar{1}$.7992	$\bar{1}$.8008	$\bar{1}$.8025	$\bar{1}$.8042	$\bar{1}$.8059	$\bar{1}$.8075	$\bar{1}$.8092	$\bar{1}$.8109	3	6	8	11	14
33	$\bar{1}$.8125	$\bar{1}$.8142	$\bar{1}$.8158	$\bar{1}$.8175	$\bar{1}$.8191	$\bar{1}$.8208	$\bar{1}$.8224	$\bar{1}$.8241	$\bar{1}$.8257	$\bar{1}$.8274	3	6	8	11	14
34	$\bar{1}$.8290	$\bar{1}$.8306	$\bar{1}$.8323	$\bar{1}$.8339	$\bar{1}$.8355	$\bar{1}$.8371	$\bar{1}$.8388	$\bar{1}$.8404	$\bar{1}$.8420	$\bar{1}$.8436	3	5	8	11	14
35	$\bar{1}$.8452	$\bar{1}$.8468	$\bar{1}$.8484	$\bar{1}$.8501	$\bar{1}$.8517	$\bar{1}$.8533	$\bar{1}$.8549	$\bar{1}$.8565	$\bar{1}$.8581	$\bar{1}$.8597	3	5	8	11	13
36	$\bar{1}$.8613	$\bar{1}$.8629	$\bar{1}$.8644	$\bar{1}$.8660	$\bar{1}$.8676	$\bar{1}$.8692	$\bar{1}$.8708	$\bar{1}$.8724	$\bar{1}$.8740	$\bar{1}$.8755	3	5	8	11	13
37	$\bar{1}$.8771	$\bar{1}$.8787	$\bar{1}$.8803	$\bar{1}$.8818	$\bar{1}$.8834	$\bar{1}$.8850	$\bar{1}$.8865	$\bar{1}$.8881	$\bar{1}$.8897	$\bar{1}$.8912	3	5	8	10	13
38	$\bar{1}$.8928	$\bar{1}$.8944	$\bar{1}$.8959	$\bar{1}$.8975	$\bar{1}$.8990	$\bar{1}$.9006	$\bar{1}$.9022	$\bar{1}$.9037	$\bar{1}$.9053	$\bar{1}$.9068	3	5	8	10	13
39	$\bar{1}$.9084	$\bar{1}$.9099	$\bar{1}$.9115	$\bar{1}$.9130	$\bar{1}$.9146	$\bar{1}$.9161	$\bar{1}$.9176	$\bar{1}$.9192	$\bar{1}$.9207	$\bar{1}$.9223	3	5	8	10	13
40	$\bar{1}$.9238	$\bar{1}$.9254	$\bar{1}$.9269	$\bar{1}$.9284	$\bar{1}$.9300	$\bar{1}$.9315	$\bar{1}$.9330	$\bar{1}$.9346	$\bar{1}$.9361	$\bar{1}$.9376	3	5	8	10	13
41	$\bar{1}$.9392	$\bar{1}$.9407	$\bar{1}$.9422	$\bar{1}$.9438	$\bar{1}$.9453	$\bar{1}$.9468	$\bar{1}$.9483	$\bar{1}$.9499	$\bar{1}$.9514	$\bar{1}$.9529	3	5	8	10	13
42	$\bar{1}$.9544	$\bar{1}$.9560	$\bar{1}$.9575	$\bar{1}$.9590	$\bar{1}$.9605	$\bar{1}$.9621	$\bar{1}$.9636	$\bar{1}$.9651	$\bar{1}$.9666	$\bar{1}$.9681	3	5	8	10	13
43	$\bar{1}$.9697	$\bar{1}$.9712	$\bar{1}$.9727	$\bar{1}$.9742	$\bar{1}$.9757	$\bar{1}$.9772	$\bar{1}$.9788	$\bar{1}$.9803	$\bar{1}$.9818	$\bar{1}$.9833	3	5	8	10	13
44	$\bar{1}$.9848	$\bar{1}$.9864	$\bar{1}$.9879	$\bar{1}$.9894	$\bar{1}$.9909	$\bar{1}$.9924	$\bar{1}$.9939	$\bar{1}$.9955	$\bar{1}$.9970	$\bar{1}$.9985	3	5	8	10	13

To find $31.23 \times \tan 28° 37'$

$$\log(31.23 \times \tan 28° 37') = \log 31.23 + \log\tan 28° 37'$$
$$= 1.4946 + \bar{1}.7369$$
$$= 1.2315$$

From the anti-log tables the answer is 17.04.

Logarithms of Tangents

°	0' 0.0°	6' 0.1°	12' 0.2°	18' 0.3°	24' 0.4°	30' 0.5°	36' 0.6°	42' 0.7°	48' 0.8°	54' 0.9°	1'	2'	3'	4'	5'
45	0.0000	0.0015	0.0030	0.0045	0.0061	0.0076	0.0091	0.0106	0.0121	0.0136	3	5	8	10	13
46	0.0152	0.0167	0.0182	0.0197	0.0212	0.0228	0.0243	0.0258	0.0273	0.0288	3	5	8	10	13
47	0.0303	0.0319	0.0334	0.0349	0.0364	0.0379	0.0395	0.0410	0.0425	0.0440	3	5	8	10	13
48	0.0456	0.0471	0.0486	0.0501	0.0517	0.0532	0.0547	0.0562	0.0578	0.0593	3	5	8	10	13
49	0.0608	0.0624	0.0639	0.0654	0.0670	0.0685	0.0700	0.0716	0.0731	0.0746	3	5	8	10	13
50	0.0762	0.0777	0.0793	0.0808	0.0824	0.0839	0.0854	0.0870	0.0885	0.0901	3	5	8	10	13
51	0.0916	0.0932	0.0947	0.0963	0.0978	0.0994	0.1010	0.1025	0.1041	0.1056	3	5	8	10	13
52	0.1072	0.1088	0.1103	0.1119	0.1135	0.1150	0.1166	0.1182	0.1197	0.1213	3	5	8	10	13
53	0.1229	0.1245	0.1260	0.1276	0.1292	0.1308	0.1324	0.1340	0.1356	0.1371	3	5	8	11	13
54	0.1387	0.1403	0.1419	0.1435	0.1451	0.1467	0.1483	0.1499	0.1516	0.1532	3	5	8	11	13
55	0.1548	0.1564	0.1580	0.1596	0.1612	0.1629	0.1645	0.1661	0.1677	0.1694	3	5	8	11	14
56	0.1710	0.1726	0.1743	0.1759	0.1776	0.1792	0.1809	0.1825	0.1842	0.1858	3	6	8	11	14
57	0.1875	0.1891	0.1908	0.1925	0.1941	0.1958	0.1975	0.1992	0.2008	0.2025	3	6	8	11	14
58	0.2042	0.2059	0.2076	0.2093	0.2110	0.2127	0.2144	0.2161	0.2178	0.2195	3	6	9	11	14
59	0.2212	0.2229	0.2247	0.2264	0.2281	0.2299	0.2316	0.2333	0.2351	0.2368	3	6	9	12	15
60	0.2386	0.2403	0.2421	0.2438	0.2456	0.2474	0.2491	0.2509	0.2527	0.2545	3	6	9	12	15
61	0.2562	0.2580	0.2598	0.2616	0.2634	0.2652	0.2670	0.2689	0.2707	0.2725	3	6	9	12	15
62	0.2743	0.2762	0.2780	0.2798	0.2817	0.2835	0.2854	0.2872	0.2891	0.2910	3	6	9	12	15
63	0.2928	0.2947	0.2966	0.2985	0.3004	0.3023	0.3042	0.3061	0.3080	0.3099	3	6	9	13	16
64	0.3118	0.3137	0.3157	0.3176	0.3196	0.3215	0.3235	0.3254	0.3274	0.3294	3	6	10	13	16
65	0.3313	0.3333	0.3353	0.3373	0.3393	0.3413	0.3433	0.3453	0.3473	0.3494	3	7	10	13	17
66	0.3514	0.3535	0.3555	0.3576	0.3596	0.3617	0.3638	0.3659	0.3679	0.3700	3	7	10	14	17
67	0.3721	0.3743	0.3764	0.3785	0.3806	0.3828	0.3849	0.3871	0.3892	0.3914	4	7	11	14	18
68	0.3936	0.3958	0.3980	0.4002	0.4024	0.4046	0.4068	0.4091	0.4113	0.4136	4	7	11	15	18
69	0.4158	0.4181	0.4204	0.4227	0.4250	0.4273	0.4296	0.4319	0.4342	0.4366	4	8	12	15	19
70	0.4389	0.4413	0.4437	0.4461	0.4484	0.4509	0.4533	0.4557	0.4581	0.4606	4	8	12	16	20
71	0.4630	0.4655	0.4680	0.4705	0.4730	0.4755	0.4780	0.4805	0.4831	0.4857	4	8	13	17	21
72	0.4882	0.4908	0.4934	0.4960	0.4986	0.5013	0.5039	0.5066	0.5093	0.5120	4	9	13	18	22
73	0.5147	0.5174	0.5201	0.5229	0.5256	0.5284	0.5312	0.5340	0.5368	0.5397	5	9	14	19	23
74	0.5425	0.5454	0.5483	0.5512	0.5541	0.5570	0.5600	0.5629	0.5659	0.5689	5	10.	15	20	24
75	0.5719	0.5750	0.5780	0.5811	0.5842	0.5873	0.5905	0.5936	0.5968	0.6000	5	10	16	21	26
76	0.6032	0.6065	0.6097	0.6130	0.6163	0.6196	0.6230	0.6264	0.6298	0.6332	6	11	17	22	28
77	0.6366	0.6401	0.6436	0.6471	0.6507	0.6542	0.6578	0.6615	0.6651	0.6688	6	12	18	24	30
78	0.6725	0.6763	0.6800	0.6838	0.6877	0.6915	0.6954	0.6994	0.7033	0.7073	6	13	19	26	32
79	0.7113	0.7154	0.7195	0.7236	0.7278	0.7320	0.7363	0.7406	0.7449	0.7493	7	14	21	28	35
80	0.7537	0.7581	0.7626	0.7672	0.7718	0.7764	0.7811	0.7858	0.7906	0.7954	8	16	23	31	39
81	0.8003	0.8052	0.8102	0.8152	0.8203	0.8255	0.8307	0.8360	0.8413	0.8467	9	17	26	35	43
82	0.8522	0.8577	0.8633	0.8690	0.8748	0.8806	0.8865	0.8924	0.8985	0.9046	10	20	29	39	49
83	0.9109	0.9172	0.9236	0.9301	0.9367	0.9433	0.9501	0.9570	0.9640	0.9711	11	22	34	45	56
84	0.9784	0.9857	0.9932	1.0008	1.0085	1.0164	1.0244	1.0326	1.0409	1.0494	13	26	40	53	66
85	1.0580	1.0669	1.0759	1.0850	1.0944	1.1040	1.1138	1.1238	1.1341	1.1446	16	32	48	64	81
86	1.1554	1.1664	1.1777	1.1893	1.2012	1.2135	1.2261	1.2391	1.2525	1.2663	Differences untrustworthy here				
87	1.2806	1.2954	1.3106	1.3264	1.3429	1.3599	1.3777	1.3962	1.4155	1.4357					
88	1.4569	1.4792	1.5027	1.5275	1.5539	1.5819	1.6119	1.6441	1.6789	1.7167					
89	1.7581	1.8038	1.8550	1.9130	1.9800	2.0591	2.1561	2.2810	2.4571	2.7581					

To find the angle A given that $\tan A = \dfrac{23.14}{17.68}$

$$\log \tan A = \log \left(\frac{23.14}{17.68}\right) = \log 23.14 - \log 17.68$$

$$= 1.3643 - 1.2475 = 0.1168$$

Using the log tangent tables

$$A = 52° 37'$$

Natural Sines

°	0' 0.0°	6' 0.1°	12' 0.2°	18' 0.3°	24' 0.4°	30' 0.5°	36' 0.6°	42' 0.7°	48' 0.8°	54' 0.9°	1'	2'	3'	4'	5
0	0.0000	0.0017	0.0035	0.0052	0.0070	0.0087	0.0105	0.0122	0.0140	0.0157	3	6	9	12	1
1	0.0175	0.0192	0.0209	0.0227	0.0244	0.0262	0.0279	0.0297	0.0314	0.0332	3	6	9	12	1
2	0.0349	0.0366	0.0384	0.0401	0.0419	0.0436	0.0454	0.0471	0.0488	0.0506	3	6	9	12	1
3	0.0523	0.0541	0.0558	0.0576	0.0593	0.0610	0.0628	0.0645	0.0663	0.0680	3	6	9	12	1
4	0.0698	0.0715	0.0732	0.0750	0.0767	0.0785	0.0802	0.0819	0.0837	0.0854	3	6	9	12	14
5	0.0872	0.0889	0.0906	0.0924	0.0941	0.0958	0.0976	0.0993	0.1011	0.1028	3	6	9	12	14
6	0.1045	0.1063	0.1080	0.1097	0.1115	0.1132	0.1149	0.1167	0.1184	0.1201	3	6	9	12	14
7	0.1219	0.1236	0.1253	0.1271	0.1288	0.1305	0.1323	0.1340	0.1357	0.1374	3	6	9	12	14
8	0.1392	0.1409	0.1426	0.1444	0.1461	0.1478	0.1495	0.1513	0.1530	0.1547	3	6	9	11	14
9	0.1564	0.1582	0.1599	0.1616	0.1633	0.1650	0.1668	0.1685	0.1702	0.1719	3	6	9	11	14
10	0.1736	0.1754	0.1771	0.1788	0.1805	0.1822	0.1840	0.1857	0.1874	0.1891	3	6	9	11	14
11	0.1908	0.1925	0.1942	0.1959	0.1977	0.1994	0.2011	0.2028	0.2045	0.2062	3	6	9	11	14
12	0.2079	0.2096	0.2113	0.2130	0.2147	0.2164	0.2181	0.2198	0.2215	0.2232	3	6	9	11	14
13	0.2250	0.2267	0.2284	0.2300	0.2317	0.2334	0.2351	0.2368	0.2385	0.2402	3	6	8	11	14
14	0.2419	0.2436	0.2453	0.2470	0.2487	0.2504	0.2521	0.2538	0.2554	0.2571	3	6	8	11	14
15	0.2588	0.2605	0.2622	0.2639	0.2656	0.2672	0.2689	0.2706	0.2723	0.2740	3	6	8	11	14
16	0.2756	0.2773	0.2790	0.2807	0.2823	0.2840	0.2857	0.2874	0.2890	0.2907	3	6	8	11	14
17	0.2924	0.2940	0.2957	0.2974	0.2990	0.3007	0.3024	0.3040	0.3057	0.3074	3	6	8	11	14
18	0.3090	0.3107	0.3123	0.3140	0.3156	0.3173	0.3190	0.3206	0.3223	0.3239	3	6	8	11	14
19	0.3256	0.3272	0.3289	0.3305	0.3322	0.3338	0.3355	0.3371	0.3387	0.3404	3	5	8	11	14
20	0.3420	0.3437	0.3453	0.3469	0.3486	0.3502	0.3518	0.3535	0.3551	0.3567	3	5	8	11	14
21	0.3584	0.3600	0.3616	0.3633	0.3649	0.3665	0.3681	0.3697	0.3714	0.3730	3	5	8	11	14
22	0.3746	0.3762	0.3778	0.3795	0.3811	0.3827	0.3843	0.3859	0.3875	0.3891	3	5	8	11	13
23	0.3907	0.3923	0.3939	0.3955	0.3971	0.3987	0.4003	0.4019	0.4035	0.4051	3	5	8	11	13
24	0.4067	0.4083	0.4099	0.4115	0.4131	0.4147	0.4163	0.4179	0.4195	0.4210	3	5	8	11	13
25	0.4226	0.4242	0.4258	0.4274	0.4289	0.4305	0.4321	0.4337	0.4352	0.4368	3	5	8	11	13
26	0.4384	0.4399	0.4415	0.4431	0.4446	0.4462	0.4478	0.4493	0.4509	0.4524	3	5	8	10	13
27	0.4540	0.4555	0.4571	0.4586	0.4602	0.4617	0.4633	0.4648	0.4664	0.4679	3	5	8	10	13
28	0.4695	0.4710	0.4726	0.4741	0.4756	0.4772	0.4787	0.4802	0.4818	0.4833	3	5	8	10	13
29	0.4848	0.4863	0.4879	0.4894	0.4909	0.4924	0.4939	0.4955	0.4970	0.4985	3	5	8	10	13
30	0.5000	0.5015	0.5030	0.5045	0.5060	0.5075	0.5090	0.5105	0.5120	0.5135	3	5	8	10	13
31	0.5150	0.5165	0.5180	0.5195	0.5210	0.5225	0.5240	0.5255	0.5270	0.5284	2	5	7	10	12
32	0.5299	0.5314	0.5329	0.5344	0.5358	0.5373	0.5388	0.5402	0.5417	0.5432	2	5	7	10	12
33	0.5446	0.5461	0.5476	0.5490	0.5505	0.5519	0.5534	0.5548	0.5563	0.5577	2	5	7	10	12
34	0.5592	0.5606	0.5621	0.5635	0.5650	0.5664	0.5678	0.5693	0.5707	0.5721	2	5	7	10	12
35	0.5736	0.5750	0.5764	0.5779	0.5793	0.5807	0.5821	0.5835	0.5850	0.5864	2	5	7	9	12
36	0.5878	0.5892	0.5906	0.5920	0.5934	0.5948	0.5962	0.5976	0.5990	0.6004	2	5	7	9	12
37	0.6018	0.6032	0.6046	0.6060	0.6074	0.6088	0.6101	0.6115	0.6129	0.6143	2	5	7	9	12
38	0.6157	0.6170	0.6184	0.6198	0.6211	0.6225	0.6239	0.6252	0.6266	0.6280	2	5	7	9	11
39	0.6293	0.6307	0.6320	0.6334	0.6347	0.6361	0.6374	0.6388	0.6401	0.6414	2	4	7	9	11
40	0.6428	0.6441	0.6455	0.6468	0.6481	0.6494	0.6508	0.6521	0.6534	0.6547	2	4	7	9	11
41	0.6561	0.6574	0.6587	0.6600	0.6613	0.6626	0.6639	0.6652	0.6665	0.6678	2	4	7	9	11
42	0.6691	0.6704	0.6717	0.6730	0.6743	0.6756	0.6769	0.6782	0.6794	0.6807	2	4	6	9	11
43	0.6820	0.6833	0.6845	0.6858	0.6871	0.6884	0.6896	0.6909	0.6921	0.6934	2	4	6	8	11
44	0.6947	0.6959	0.6972	0.6984	0.6997	0.7009	0.7022	0.7034	0.7046	0.7059	2	4	6	8	10

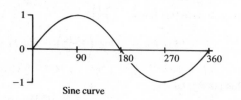

Sine curve

Natural Sines

0' 0.0°	6' 0.1°	12' 0.2°	18' 0.3°	24' 0.4°	30' 0.5°	36' 0.6°	42' 0.7°	48' 0.8°	54' 0.9°	1'	2'	3'	4'	5'
0.7071	0.7083	0.7096	0.7108	0.7120	0.7133	0.7145	0.7157	0.7169	0.7181	2	4	6	8	10
0.7193	0.7206	0.7218	0.7230	0.7242	0.7254	0.7266	0.7278	0.7290	0.7302	2	4	6	8	10
0.7314	0.7325	0.7337	0.7349	0.7361	0.7373	0.7385	0.7396	0.7408	0.7420	2	4	6	8	10
0.7431	0.7443	0.7455	0.7466	0.7478	0.7490	0.7501	0.7513	0.7524	0.7536	2	4	6	8	10
0.7547	0.7558	0.7570	0.7581	0.7593	0.7604	0.7615	0.7627	0.7638	0.7649	2	4	6	8	9
0.7660	0.7672	0.7683	0.7694	0.7705	0.7716	0.7727	0.7738	0.7749	0.7760	2	4	6	7	9
0.7771	0.7782	0.7793	0.7804	0.7815	0.7826	0.7837	0.7848	0.7859	0.7869	2	4	5	7	9
0.7880	0.7891	0.7902	0.7912	0.7923	0.7934	0.7944	0.7955	0.7965	0.7976	2	4	5	7	9
0.7986	0.7997	0.8007	0.8018	0.8028	0.8039	0.8049	0.8059	0.8070	0.8080	2	3	5	7	9
0.8090	0.8100	0.8111	0.8121	0.8131	0.8141	0.8151	0.8161	0.8171	0.8181	2	3	5	7	8
0.8192	0.8202	0.8211	0.8221	0.8231	0.8241	0.8251	0.8261	0.8271	0.8281	2	3	5	7	8
0.8290	0.8300	0.8310	0.8320	0.8329	0.8339	0.8348	0.8358	0.8368	0.8377	2	3	5	6	8
0.8387	0.8396	0.8406	0.8415	0.8425	0.8434	0.8443	0.8453	0.8462	0.8471	2	3	5	6	8
0.8480	0.8490	0.8499	0.8508	0.8517	0.8526	0.8536	0.8545	0.8554	0.8563	2	3	5	6	8
0.8572	0.8581	0.8590	0.8599	0.8607	0.8616	0.8625	0.8634	0.8643	0.8652	1	3	4	6	7
0.8660	0.8669	0.8678	0.8686	0.8695	0.8704	0.8712	0.8721	0.8729	0.8738	1	3	4	6	7
0.8746	0.8755	0.8763	0.8771	0.8780	0.8788	0.8796	0.8805	0.8813	0.8821	1	3	4	6	7
0.8829	0.8838	0.8846	0.8854	0.8862	0.8870	0.8878	0.8886	0.8894	0.8902	1	3	4	5	7
0.8910	0.8918	0.8926	0.8934	0.8942	0.8949	0.8957	0.8965	0.8973	0.8980	1	3	4	5	6
0.8988	0.8996	0.9003	0.9011	0.9018	0.9026	0.9033	0.9041	0.9048	0.9056	1	3	4	5	6
0.9063	0.9070	0.9078	0.9085	0.9092	0.9100	0.9107	0.9114	0.9121	0.9128	1	2	4	5	6
0.9135	0.9143	0.9150	0.9157	0.9164	0.9171	0.9178	0.9184	0.9191	0.9198	1	2	3	5	6
0.9205	0.9212	0.9219	0.9225	0.9232	0.9239	0.9245	0.9252	0.9259	0.9265	1	2	3	4	6
0.9272	0.9278	0.9285	0.9291	0.9298	0.9304	0.9311	0.9317	0.9323	0.9330	1	2	3	4	5
0.9336	0.9342	0.9348	0.9354	0.9361	0.9367	0.9373	0.9379	0.9385	0.9391	1	2	3	4	5
0.9397	0.9403	0.9409	0.9415	0.9421	0.9426	0.9432	0.9438	0.9444	0.9449	1	2	3	4	5
0.9455	0.9461	0.9466	0.9472	0.9478	0.9483	0.9489	0.9494	0.9500	0.9505	1	2	3	4	5
0.9511	0.9516	0.9521	0.9527	0.9532	0.9537	0.9542	0.9548	0.9553	0.9558	1	2	3	3	4
0.9563	0.9568	0.9573	0.9578	0.9583	0.9588	0.9593	0.9598	0.9603	0.9608	1	2	2	3	4
0.9613	0.9617	0.9622	0.9627	0.9632	0.9636	0.9641	0.9646	0.9650	0.9655	1	2	2	3	4
0.9659	0.9664	0.9668	0.9673	0.9677	0.9681	0.9686	0.9690	0.9694	0.9699	1	1	2	3	4
0.9703	0.9707	0.9711	0.9715	0.9720	0.9724	0.9728	0.9732	0.9736	0.9740	1	1	2	3	3
0.9744	0.9748	0.9751	0.9755	0.9759	0.9763	0.9767	0.9770	0.9774	0.9778	1	1	2	2	3
0.9781	0.9785	0.9789	0.9792	0.9796	0.9799	0.9803	0.9806	0.9810	0.9813	1	1	2	2	3
0.9816	0.9820	0.9823	0.9826	0.9829	0.9833	0.9836	0.9839	0.9842	0.9845	1	1	2	2	3
0.9848	0.9851	0.9854	0.9857	0.9860	0.9863	0.9866	0.9869	0.9871	0.9874	0	1	1	2	2
0.9877	0.9880	0.9882	0.9885	0.9888	0.9890	0.9893	0.9895	0.9898	0.9900	0	1	1	2	2
0.9903	0.9905	0.9907	0.9910	0.9912	0.9914	0.9917	0.9919	0.9921	0.9923	0	1	1	1	2
0.9925	0.9928	0.9930	0.9932	0.9934	0.9936	0.9938	0.9940	0.9942	0.9943	0	1	1	1	2
0.9945	0.9947	0.9949	0.9951	0.9952	0.9954	0.9956	0.9957	0.9959	0.9960	0	1	1	1	1
0.9962	0.9963	0.9965	0.9966	0.9968	0.9969	0.9971	0.9972	0.9973	0.9974	0	0	1	1	1
0.9976	0.9977	0.9978	0.9979	0.9980	0.9981	0.9982	0.9983	0.9984	0.9985	0	0	1	1	1
0.9986	0.9987	0.9988	0.9989	0.9990	0.9990	0.9991	0.9992	0.9993	0.9993	0	0	0	1	1
0.9994	0.9995	0.9995	0.9996	0.9996	0.9997	0.9997	0.9997	0.9998	0.9998	0	0	0	0	0
0.9998	0.9999	0.9999	0.9999	0.9999	1.0000	1.0000	1.0000	1.0000	1.0000	0	0	0	0	0
1.0000														

Quadrant	Angle	sin A =	Examples
first	0 to 90°	sin A	sin 34°38′ = 0.5683
second	90° to 180°	sin(180° − A)	sin 145°22′ = sin(180° − 145°22′)
third	180° to 270°	−sin(A − 180°)	= sin 34°38′ = 0.5683
fourth	270° to 360°	−sin(360° − A)	sin 214°38′ = −sin(214°38′ − 180°)
			= −sin 34°38′ = −0.5683
			sin 325°22′ = −sin(360° − 325°22′)
			= −sin 34°38′ = −0.5683

Natural Cosines

Numbers in difference columns to be *subtracted*, not added.

°	0' 0.0°	6' 0.1°	12' 0.2°	18' 0.3°	24' 0.4°	30' 0.5°	36' 0.6°	42' 0.7°	48' 0.8°	54' 0.9°	1'	2'	3'	4'	
0	1.0000	1.0000	1.0000	1.0000	1.0000	1.0000	0.9999	0.9999	0.9999	0.9999	0	0	0	0	
1	0.9998	0.9998	0.9998	0.9997	0.9997	0.9997	0.9996	0.9996	0.9995	0.9995	0	0	0	0	
2	0.9994	0.9993	0.9993	0.9992	0.9991	0.9990	0.9990	0.9989	0.9988	0.9987	0	0	0	1	
3	0.9986	0.9985	0.9984	0.9983	0.9982	0.9981	0.9980	0.9979	0.9978	0.9977	0	0	1	1	
4	0.9976	0.9974	0.9973	0.9972	0.9971	0.9969	0.9968	0.9966	0.9965	0.9963	0	0	1	1	
5	0.9962	0.9960	0.9959	0.9957	0.9956	0.9954	0.9952	0.9951	0.9949	0.9947	0	1	1	1	
6	0.9945	0.9943	0.9942	0.9940	0.9938	0.9936	0.9934	0.9932	0.9930	0.9928	0	1	1	1	
7	0.9925	0.9923	0.9921	0.9919	0.9917	0.9914	0.9912	0.9910	0.9907	0.9905	0	1	1	1	
8	0.9903	0.9900	0.9898	0.9895	0.9893	0.9890	0.9888	0.9885	0.9882	0.9880	0	1	1	2	
9	0.9877	0.9874	0.9871	0.9869	0.9866	0.9863	0.9860	0.9857	0.9854	0.9851	0	1	1	2	
10	0.9848	0.9845	0.9842	0.9839	0.9836	0.9833	0.9829	0.9826	0.9823	0.9820	1	1	2	2	
11	0.9816	0.9813	0.9810	0.9806	0.9803	0.9799	0.9796	0.9792	0.9789	0.9785	1	1	2	2	
12	0.9781	0.9778	0.9774	0.9770	0.9767	0.9763	0.9759	0.9755	0.9751	0.9748	1	1	2	2	
13	0.9744	0.9740	0.9736	0.9732	0.9728	0.9724	0.9720	0.9715	0.9711	0.9707	1	1	2	3	
14	0.9703	0.9699	0.9694	0.9690	0.9686	0.9681	0.9677	0.9673	0.9668	0.9664	1	1	2	3	
15	0.9659	0.9655	0.9650	0.9646	0.9641	0.9636	0.9632	0.9627	0.9622	0.9617	1	2	2	3	
16	0.9613	0.9608	0.9603	0.9598	0.9593	0.9588	0.9583	0.9578	0.9573	0.9568	1	2	2	3	
17	0.9563	0.9558	0.9553	0.9548	0.9542	0.9537	0.9532	0.9527	0.9521	0.9516	1	2	3	3	
18	0.9511	0.9505	0.9500	0.9494	0.9489	0.9483	0.9478	0.9472	0.9466	0.9461	1	2	3	4	
19	0.9455	0.9449	0.9444	0.9438	0.9432	0.9426	0.9421	0.9415	0.9409	0.9403	1	2	3	4	
20	0.9397	0.9391	0.9385	0.9379	0.9373	0.9367	0.9361	0.9354	0.9348	0.9342	1	2	3	4	
21	0.9336	0.9330	0.9323	0.9317	0.9311	0.9304	0.9298	0.9291	0.9285	0.9278	1	2	3	4	
22	0.9272	0.9265	0.9259	0.9252	0.9245	0.9239	0.9232	0.9225	0.9219	0.9212	1	2	3	4	
23	0.9205	0.9198	0.9191	0.9184	0.9178	0.9171	0.9164	0.9157	0.9150	0.9143	1	2	3	5	
24	0.9135	0.9128	0.9121	0.9114	0.9107	0.9100	0.9092	0.9085	0.9078	0.9070	1	2	4	5	
25	0.9063	0.9056	0.9048	0.9041	0.9033	0.9026	0.9018	0.9011	0.9003	0.8996	1	3	4	5	
26	0.8988	0.8980	0.8973	0.8965	0.8957	0.8949	0.8942	0.8934	0.8926	0.8918	1	3	4	5	
27	0.8910	0.8902	0.8894	0.8886	0.8878	0.8870	0.8862	0.8854	0.8846	0.8838	1	3	4	5	
28	0.8829	0.8821	0.8813	0.8805	0.8796	0.8788	0.8780	0.8771	0.8763	0.8755	1	3	4	6	
29	0.8746	0.8738	0.8729	0.8721	0.8712	0.8704	0.8695	0.8686	0.8678	0.8669	1	3	4	6	
30	0.8660	0.8652	0.8643	0.8634	0.8625	0.8616	0.8607	0.8599	0.8590	0.8581	1	3	4	6	
31	0.8572	0.8563	0.8554	0.8545	0.8536	0.8526	0.8517	0.8508	0.8499	0.8490	2	3	5	6	
32	0.8480	0.8471	0.8462	0.8453	0.8443	0.8434	0.8425	0.8415	0.8406	0.8396	2	3	5	6	
33	0.8387	0.8377	0.8368	0.8358	0.8348	0.8339	0.8329	0.8320	0.8310	0.8300	2	3	5	6	
34	0.8290	0.8281	0.8271	0.8261	0.8251	0.8241	0.8231	0.8221	0.8211	0.8202	2	3	5	7	
35	0.8192	0.8181	0.8171	0.8161	0.8151	0.8141	0.8131	0.8121	0.8111	0.8100	2	3	5	7	
36	0.8090	0.8080	0.8070	0.8059	0.8049	0.8039	0.8028	0.8018	0.8007	0.7997	2	3	5	7	
37	0.7986	0.7976	0.7965	0.7955	0.7944	0.7934	0.7923	0.7912	0.7902	0.7891	2	4	5	7	
38	0.7880	0.7869	0.7859	0.7848	0.7837	0.7826	0.7815	0.7804	0.7793	0.7782	2	4	5	7	
39	0.7771	0.7760	0.7749	0.7738	0.7727	0.7716	0.7705	0.7694	0.7683	0.7672	2	4	6	7	
40	0.7660	0.7649	0.7638	0.7627	0.7615	0.7604	0.7593	0.7581	0.7570	0.7559	2	4	6	8	
41	0.7547	0.7536	0.7524	0.7513	0.7501	0.7490	0.7478	0.7466	0.7455	0.7443	2	4	6	8	1
42	0.7431	0.7420	0.7408	0.7396	0.7385	0.7373	0.7361	0.7349	0.7337	0.7325	2	4	6	8	1
43	0.7314	0.7302	0.7290	0.7278	0.7266	0.7254	0.7242	0.7230	0.7218	0.7206	2	4	6	8	1
44	0.7193	0.7181	0.7169	0.7157	0.7145	0.7133	0.7120	0.7108	0.7096	0.7083	2	4	6	8	1

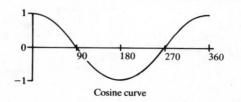

Cosine curve

56

Natural Cosines

Numbers in difference columns to be *subtracted*, not added.

°	0' 0.0°	6' 0.1°	12' 0.2°	18' 0.3°	24' 0.4°	30' 0.5°	36' 0.6°	42' 0.7°	48' 0.8°	54' 0.9°	1'	2'	3'	4'	5'
45	0.7071	0.7059	0.7046	0.7034	0.7022	0.7009	0.6997	0.6984	0.6972	0.6959	2	4	6	8	10
46	0.6947	0.6934	0.6921	0.6909	0.6896	0.6884	0.6871	0.6858	0.6845	0.6833	2	4	6	8	11
47	0.6820	0.6807	0.6794	0.6782	0.6769	0.6756	0.6743	0.6730	0.6717	0.6704	2	4	6	9	11
48	0.6691	0.6678	0.6665	0.6652	0.6639	0.6626	0.6613	0.6600	0.6587	0.6574	2	4	7	9	11
49	0.6561	0.6547	0.6534	0.6521	0.6508	0.6494	0.6481	0.6468	0.6455	0.6441	2	4	7	9	11
50	0.6428	0.6414	0.6401	0.6388	0.6374	0.6361	0.6347	0.6334	0.6320	0.6307	2	4	7	9	11
51	0.6293	0.6280	0.6266	0.6252	0.6239	0.6225	0.6211	0.6198	0.6184	0.6170	2	5	7	9	11
52	0.6157	0.6143	0.6129	0.6115	0.6101	0.6088	0.6074	0.6060	0.6046	0.6032	2	5	7	9	12
53	0.6018	0.6004	0.5990	0.5976	0.5962	0.5948	0.5934	0.5920	0.5906	0.5892	2	5	7	9	12
54	0.5878	0.5864	0.5850	0.5835	0.5821	0.5807	0.5793	0.5779	0.5764	0.5750	2	5	7	9	12
55	0.5736	0.5721	0.5707	0.5693	0.5678	0.5664	0.5650	0.5635	0.5621	0.5606	2	5	7	10	12
56	0.5592	0.5577	0.5563	0.5548	0.5534	0.5519	0.5505	0.5490	0.5476	0.5461	2	5	7	10	12
57	0.5446	0.5432	0.5417	0.5402	0.5388	0.5373	0.5358	0.5344	0.5329	0.5314	2	5	7	10	12
58	0.5299	0.5284	0.5270	0.5255	0.5240	0.5225	0.5210	0.5195	0.5180	0.5165	2	5	7	10	12
59	0.5150	0.5135	0.5120	0.5105	0.5090	0.5075	0.5060	0.5045	0.5030	0.5015	3	5	8	10	13
60	0.5000	0.4985	0.4970	0.4955	0.4939	0.4924	0.4909	0.4894	0.4879	0.4863	3	5	8	10	13
61	0.4848	0.4833	0.4818	0.4802	0.4787	0.4772	0.4756	0.4741	0.4726	0.4710	3	5	8	10	13
62	0.4695	0.4679	0.4664	0.4648	0.4633	0.4617	0.4602	0.4586	0.4571	0.4555	3	5	8	10	13
63	0.4540	0.4524	0.4509	0.4493	0.4478	0.4462	0.4446	0.4431	0.4415	0.4399	3	5	8	10	13
64	0.4384	0.4368	0.4352	0.4337	0.4321	0.4305	0.4289	0.4274	0.4258	0.4242	3	5	8	11	13
65	0.4226	0.4210	0.4195	0.4179	0.4163	0.4147	0.4131	0.4115	0.4099	0.4083	3	5	8	11	13
66	0.4067	0.4051	0.4035	0.4019	0.4003	0.3987	0.3971	0.3955	0.3939	0.3923	3	5	8	11	13
67	0.3907	0.3891	0.3875	0.3859	0.3843	0.3827	0.3811	0.3795	0.3778	0.3762	3	5	8	11	13
68	0.3746	0.3730	0.3714	0.3697	0.3681	0.3665	0.3649	0.3633	0.3616	0.3600	3	5	8	11	14
69	0.3584	0.3567	0.3551	0.3535	0.3518	0.3502	0.3486	0.3469	0.3453	0.3437	3	5	8	11	14
70	0.3420	0.3404	0.3387	0.3371	0.3355	0.3338	0.3322	0.3305	0.3289	0.3272	3	5	8	11	14
71	0.3256	0.3239	0.3223	0.3206	0.3190	0.3173	0.3156	0.3140	0.3123	0.3107	3	6	8	11	14
72	0.3090	0.3074	0.3057	0.3040	0.3024	0.3007	0.2990	0.2974	0.2957	0.2940	3	6	8	11	14
73	0.2924	0.2907	0.2890	0.2874	0.2857	0.2840	0.2823	0.2807	0.2790	0.2773	3	6	8	11	14
74	0.2756	0.2740	0.2723	0.2706	0.2689	0.2672	0.2656	0.2639	0.2622	0.2605	3	6	8	11	14
75	0.2588	0.2571	0.2554	0.2538	0.2521	0.2504	0.2487	0.2470	0.2453	0.2436	3	6	8	11	14
76	0.2419	0.2402	0.2385	0.2368	0.2351	0.2334	0.2317	0.2300	0.2284	0.2267	3	6	8	11	14
77	0.2250	0.2233	0.2215	0.2198	0.2181	0.2164	0.2147	0.2130	0.2113	0.2096	3	6	9	11	14
78	0.2079	0.2062	0.2045	0.2028	0.2011	0.1994	0.1977	0.1959	0.1942	0.1925	3	6	9	11	14
79	0.1908	0.1891	0.1874	0.1857	0.1840	0.1822	0.1805	0.1788	0.1771	0.1754	3	6	9	11	14
80	0.1736	0.1719	0.1702	0.1685	0.1668	0.1650	0.1633	0.1616	0.1599	0.1582	3	6	9	11	14
81	0.1554	0.1547	0.1530	0.1513	0.1495	0.1478	0.1461	0.1444	0.1426	0.1409	3	6	9	11	14
82	0.1392	0.1374	0.1357	0.1340	0.1323	0.1305	0.1288	0.1271	0.1253	0.1236	3	6	9	12	14
83	0.1219	0.1201	0.1184	0.1167	0.1149	0.1132	0.1115	0.1097	0.1080	0.1063	3	6	9	12	14
84	0.1045	0.1028	0.1011	0.0993	0.0976	0.0958	0.0941	0.0924	0.0906	0.0889	3	6	9	12	14
85	0.0872	0.0854	0.0837	0.0819	0.0802	0.0785	0.0767	0.0750	0.0732	0.0715	3	6	9	12	14
86	0.0698	0.0680	0.0663	0.0645	0.0628	0.0610	0.0593	0.0576	0.0558	0.0541	3	6	9	12	15
87	0.0523	0.0506	0.0488	0.0471	0.0454	0.0436	0.0419	0.0401	0.0384	0.0366	3	6	9	12	15
88	0.0349	0.0332	0.0314	0.0297	0.0279	0.0262	0.0244	0.0227	0.0209	0.0192	3	6	9	12	15
89	0.0175	0.0157	0.0140	0.0122	0.0105	0.0087	0.0070	0.0052	0.0035	0.0017	3	6	9	12	15
90	0.0000														

Quadrant	Angle	cos A =	Examples
first	0 to 90°	cos A	cos 33°26' = 0.8345
second	90° to 180°	−cos(180° − A)	cos 146°34' = −cos(180° − 146°34')
third	180° to 270°	−cos(A − 180°)	= −cos 33°26' = −0.8345
fourth	270° to 360°	cos(360° − A)	cos 213°26' = −cos(213°26' − 180°)
			= −cos 33°26' = −0.8345
			cos 326°34' = cos(360° − 326°34')
			= cos 33°26' = 0.8345

Natural Tangents

°	0' 0.0°	6' 0.1°	12' 0.2°	18' 0.3°	24' 0.4°	30' 0.5°	36' 0.6°	42' 0.7°	48' 0.8°	54' 0.9°	1'	2'	3'	4'	5'
0	0.0000	0.0017	0.0035	0.0052	0.0070	0.0087	0.0105	0.0122	0.0140	0.0157	3	6	9	12	15
1	0.0175	0.0192	0.0209	0.0227	0.0244	0.0262	0.0279	0.0297	0.0314	0.0332	3	6	9	12	15
2	0.0349	0.0367	0.0384	0.0402	0.0419	0.0437	0.0454	0.0472	0.0489	0.0507	3	6	9	12	15
3	0.0524	0.0542	0.0559	0.0577	0.0594	0.0612	0.0629	0.0647	0.0664	0.0682	3	6	9	12	15
4	0.0699	0.0717	0.0734	0.0752	0.0769	0.0787	0.0805	0.0822	0.0840	0.0857	3	6	9	12	15
5	0.0875	0.0892	0.0910	0.0928	0.0945	0.0963	0.0981	0.0998	0.1016	0.1033	3	6	9	12	15
6	0.1051	0.1069	0.1086	0.1104	0.1122	0.1139	0.1157	0.1175	0.1192	0.1210	3	6	9	12	15
7	0.1228	0.1246	0.1263	0.1281	0.1299	0.1317	0.1334	0.1352	0.1370	0.1388	3	6	9	12	15
8	0.1405	0.1423	0.1441	0.1459	0.1477	0.1495	0.1512	0.1530	0.1548	0.1566	3	6	9	12	15
9	0.1584	0.1602	0.1620	0.1638	0.1655	0.1673	0.1691	0.1709	0.1727	0.1745	3	6	9	12	15
10	0.1763	0.1781	0.1799	0.1817	0.1835	0.1853	0.1871	0.1890	0.1908	0.1926	3	6	9	12	15
11	0.1944	0.1962	0.1980	0.1998	0.2016	0.2035	0.2053	0.2071	0.2089	0.2107	3	6	9	12	15
12	0.2126	0.2144	0.2162	0.2180	0.2199	0.2217	0.2235	0.2254	0.2272	0.2290	3	6	9	12	15
13	0.2309	0.2327	0.2345	0.2364	0.2382	0.2401	0.2419	0.2438	0.2456	0.2475	3	6	9	12	15
14	0.2493	0.2512	0.2530	0.2549	0.2568	0.2586	0.2605	0.2623	0.2642	0.2661	3	6	9	12	16
15	0.2679	0.2698	0.2717	0.2736	0.2754	0.2773	0.2792	0.2811	0.2830	0.2849	3	6	9	13	16
16	0.2867	0.2886	0.2905	0.2924	0.2943	0.2962	0.2981	0.3000	0.3019	0.3038	3	6	9	13	16
17	0.3057	0.3076	0.3096	0.3115	0.3134	0.3153	0.3172	0.3191	0.3211	0.3230	3	6	10	13	16
18	0.3249	0.3269	0.3288	0.3307	0.3327	0.3346	0.3365	0.3385	0.3404	0.3424	3	6	10	13	16
19	0.3443	0.3463	0.3482	0.3502	0.3522	0.3541	0.3561	0.3581	0.3600	0.3620	3	7	10	13	16
20	0.3640	0.3659	0.3679	0.3699	0.3719	0.3739	0.3759	0.3779	0.3799	0.3819	3	7	10	13	17
21	0.3839	0.3859	0.3879	0.3899	0.3919	0.3939	0.3959	0.3979	0.4000	0.4020	3	7	10	13	17
22	0.4040	0.4061	0.4081	0.4101	0.4122	0.4142	0.4163	0.4183	0.4204	0.4224	3	7	10	14	17
23	0.4245	0.4265	0.4286	0.4307	0.4327	0.4348	0.4369	0.4390	0.4411	0.4431	3	7	10	14	17
24	0.4452	0.4473	0.4494	0.4515	0.4536	0.4557	0.4578	0.4599	0.4621	0.4642	4	7	11	14	18
25	0.4663	0.4684	0.4706	0.4727	0.4748	0.4770	0.4791	0.4813	0.4834	0.4856	4	7	11	14	18
26	0.4877	0.4899	0.4921	0.4942	0.4964	0.4986	0.5008	0.5029	0.5051	0.5073	4	7	11	15	18
27	0.5095	0.5117	0.5139	0.5161	0.5184	0.5206	0.5228	0.5250	0.5272	0.5295	4	7	11	15	18
28	0.5317	0.5340	0.5362	0.5384	0.5407	0.5430	0.5452	0.5475	0.5498	0.5520	4	8	11	15	19
29	0.5543	0.5566	0.5589	0.5612	0.5635	0.5658	0.5681	0.5704	0.5727	0.5750	4	8	12	15	19
30	0.5774	0.5797	0.5820	0.5844	0.5867	0.5890	0.5914	0.5938	0.5961	0.5985	4	8	12	16	20
31	0.6009	0.6032	0.6056	0.6080	0.6104	0.6128	0.6152	0.6176	0.6200	0.6224	4	8	12	16	20
32	0.6249	0.6273	0.6297	0.6322	0.6346	0.6371	0.6395	0.6420	0.6445	0.6469	4	8	12	16	20
33	0.6494	0.6519	0.6544	0.6569	0.6594	0.6619	0.6644	0.6669	0.6694	0.6720	4	8	13	17	21
34	0.6745	0.6771	0.6796	0.6822	0.6847	0.6873	0.6899	0.6924	0.6950	0.6976	4	9	13	17	21
35	0.7002	0.7028	0.7054	0.7080	0.7107	0.7133	0.7159	0.7186	0.7212	0.7239	4	9	13	17	22
36	0.7265	0.7292	0.7319	0.7346	0.7373	0.7400	0.7427	0.7454	0.7481	0.7508	5	9	14	18	23
37	0.7536	0.7563	0.7590	0.7618	0.7646	0.7673	0.7701	0.7729	0.7757	0.7785	5	9	14	18	23
38	0.7813	0.7841	0.7869	0.7898	0.7926	0.7954	0.7983	0.8012	0.8040	0.8069	5	9	14	19	24
39	0.8098	0.8127	0.8156	0.8185	0.8214	0.8243	0.8273	0.8302	0.8332	0.8361	5	10	15	20	24
40	0.8391	0.8421	0.8451	0.8481	0.8511	0.8541	0.8571	0.8601	0.8632	0.8662	5	10	15	20	25
41	0.8693	0.8724	0.8754	0.8785	0.8816	0.8847	0.8878	0.8910	0.8941	0.8972	5	10	16	21	26
42	0.9004	0.9036	0.9067	0.9099	0.9131	0.9163	0.9195	0.9228	0.9260	0.9293	5	11	16	21	27
43	0.9325	0.9358	0.9391	0.9424	0.9457	0.9490	0.9523	0.9556	0.9590	0.9623	6	11	17	22	28
44	0.9657	0.9691	0.9725	0.9759	0.9793	0.9827	0.9861	0.9896	0.9930	0.9965	6	11	17	23	28

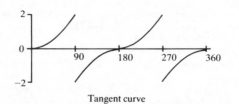

Tangent curve

Natural Tangents

°	0' 0.0°	6' 0.1°	12' 0.2°	18' 0.3°	24' 0.4°	30' 0.5°	36' 0.6°	42' 0.7°	48' 0.8°	54' 0.9°	1'	2'	3'	4'	5'
45	1.0000	1.0035	1.0070	1.0105	1.0141	1.0176	1.0212	1.0247	1.0283	1.0319	6	12	18	24	30
46	1.0355	1.0392	1.0428	1.0464	1.0501	1.0538	1.0575	1.0612	1.0649	1.0686	6	12	18	25	31
47	1.0724	1.0761	1.0799	1.0837	1.0875	1.0913	1.0951	1.0990	1.1028	1.1067	6	13	19	25	32
48	1.1106	1.1145	1.1184	1.1224	1.1263	1.1303	1.1343	1.1383	1.1423	1.1463	7	13	20	27	33
49	1.1504	1.1544	1.1585	1.1626	1.1667	1.1708	1.1750	1.1792	1.1833	1.1875	7	14	21	28	34
50	1.1918	1.1960	1.2002	1.2045	1.2088	1.2131	1.2174	1.2218	1.2261	1.2305	7	14	22	29	36
51	1.2349	1.2393	1.2437	1.2482	1.2527	1.2572	1.2617	1.2662	1.2708	1.2753	8	15	23	30	38
52	1.2799	1.2846	1.2892	1.2938	1.2985	1.3032	1.3079	1.3127	1.3175	1.3222	8	16	24	31	39
53	1.3270	1.3319	1.3367	1.3416	1.3465	1.3514	1.3564	1.3613	1.3663	1.3713	8	16	25	33	41
54	1.3764	1.3814	1.3865	1.3916	1.3968	1.4019	1.4071	1.4124	1.4176	1.4229	9	17	26	34	43
55	1.4281	1.4335	1.4388	1.4442	1.4496	1.4550	1.4605	1.4659	1.4715	1.4770	9	18	27	36	45
56	1.4826	1.4882	1.4938	1.4994	1.5051	1.5108	1.5166	1.5224	1.5282	1.5340	10	19	29	38	48
57	1.5399	1.5458	1.5517	1.5577	1.5637	1.5697	1.5757	1.5818	1.5880	1.5941	10	20	30	40	50
58	1.6003	1.6066	1.6128	1.6191	1.6255	1.6319	1.6383	1.6447	1.6512	1.6577	11	21	32	43	53
59	1.6643	1.6709	1.6775	1.6842	1.6909	1.6977	1.7045	1.7113	1.7182	1.7251	11	23	34	45	56
60	1.7321	1.7391	1.7461	1.7532	1.7603	1.7675	1.7747	1.7820	1.7893	1.7966	12	24	36	48	60
61	1.8040	1.8115	1.8190	1.8265	1.8341	1.8418	1.8495	1.8572	1.8650	1.8728	13	26	38	51	64
62	1.8807	1.8887	1.8967	1.9047	1.9128	1.9210	1.9292	1.9375	1.9458	1.9542	14	27	41	55	68
63	1.9626	1.9711	1.9797	1.9883	1.9970	2.0057	2.0145	2.0233	2.0323	2.0413	15	29	44	58	73
64	2.0503	2.0594	2.0686	2.0778	2.0872	2.0965	2.1060	2.1155	2.1251	2.1348	16	31	47	63	78
65	2.1445	2.1543	2.1642	2.1742	2.1842	2.1943	2.2045	2.2148	2.2251	2.2355	17	34	51	68	85
66	2.2460	2.2566	2.2673	2.2781	2.2889	2.2998	2.3109	2.3220	2.3332	2.3445	18	37	55	73	92
67	2.3559	2.3673	2.3789	2.3906	2.4023	2.4142	2.4262	2.4383	2.4504	2.4627	20	40	60	79	99
68	2.4751	2.4876	2.5002	2.5129	2.5257	2.5386	2.5517	2.5649	2.5782	2.5916	22	43	65	87	108
69	2.6051	2.6187	2.6325	2.6464	2.6605	2.6746	2.6889	2.7034	2.7179	2.7326	24	47	71	95	119
70	2.7475	2.7625	2.7776	2.7929	2.8083	2.8329	2.8397	2.8556	2.8716	2.8878	26	52	78	104	131
71	2.9042	2.9208	2.9375	2.9544	2.9714	2.9887	3.0061	3.0237	3.0415	3.0595	29	58	87	116	145
72	3.0777	3.0961	3.1146	3.1334	3.1524	3.1716	3.1910	3.2106	3.2305	3.2506	32	64	96	129	161
73	3.2709	3.2914	3.3122	3.3332	3.3544	3.3759	3.3977	3.4197	3.4420	3.4646	36	72	108	144	180
74	3.4874	3.5105	3.5339	3.5576	3.5816	3.6059	3.6305	3.6554	3.6806	3.7062	41	81	122	163	204
75	3.7321	3.7583	3.7848	3.8118	3.8391	3.8667	3.8947	3.9232	3.9520	3.9812	46	93	139	186	232
76	4.0108	4.0408	4.0713	4.1022	4.1335	4.1653	4.1976	4.2303	4.2635	4.2972	53	107	160	213	267
77	4.3315	4.3662	4.4015	4.4374	4.4737	4.5107	4.5483	4.5864	4.6252	4.6646					
78	4.7046	4.7453	4.7867	4.8288	4.8716	4.9152	4.9594	5.0045	5.0504	5.0970					
79	5.1446	5.1929	5.2422	5.2924	5.3435	5.3955	5.4486	5.5026	5.5578	5.6140					
80	5.6713	5.7297	5.7894	5.8502	5.9124	5.9758	6.0405	6.1066	6.1742	6.2432	Differences				
81	6.3138	6.3859	6.4596	6.5350	6.6122	6.6912	6.7720	6.8548	6.9395	7.0264	untrustworthy				
82	7.1154	7.2066	7.3002	7.3962	7.4947	7.5958	7.6996	7.8062	7.9158	8.0285	here				
83	8.1443	8.2636	8.3863	8.5126	8.6427	8.7769	8.9152	9.0579	9.2052	9.3572					
84	9.5144	9.677	9.845	10.02	10.20	10.39	10.58	10.78	10.99	11.20					
85	11.43	11.66	11.91	12.16	12.43	12.71	13.00	13.30	13.62	13.95					
86	14.30	14.67	15.06	15.46	15.89	16.35	16.83	17.34	17.89	18.46					
87	19.08	19.74	20.45	21.20	22.02	22.90	23.86	24.90	26.03	27.27					
88	28.64	30.14	31.82	33.69	35.80	38.19	40.92	44.07	47.74	52.08					
89	57.29	63.66	71.62	81.85	95.49	114.6	143.2	191.0	286.5	57.30					
90	∞														

Quadrant	Angle	tan A =	Examples
first	0 to 90°	tan A	tan 56°17′ = 1.4986
second	90° to 180°	−tan(180° − A)	tan 123°43′ = −tan(180° − 123°43′)
third	180° to 270°	tan(A − 180°)	= −tan 56°17′ = −1.4986
fourth	270° to 360°	−tan(360° − A)	tan 236°17′ = tan(236°17′ − 180°)
			= tan 56°17′ = 1.4986
			tan 303°43′ = −tan(360° − 303°43′)
			= −tan 56°17′ = −1.4986

Natural Cotangents

Numbers in difference columns to be *subtracted*, not added.

°	0' 0.0°	6' 0.1°	12' 0.2°	18' 0.3°	24' 0.4°	30' 0.5°	36' 0.6°	42' 0.7°	48' 0.8°	54' 0.9°	1'	2'	3'	4'	5'
0	∞	573.0	286.5	191.0	143.2	114.6	95.49	81.85	71.62	63.66					
1	57.29	52.08	47.74	44.07	40.92	38.19	35.80	33.69	31.82	30.14		Differences			
2	28.64	27.27	26.03	24.90	23.86	22.90	22.02	21.20	20.45	19.74		untrustworthy			
3	19.08	18.46	17.89	17.34	16.83	16.35	15.89	15.46	15.06	14.67		here			
4	14.30	13.95	13.62	13.30	13.00	12.71	12.43	12.16	11.91	11.66					
5	11.43	11.20	10.99	10.78	10.58	10.39	10.20	10.02	9.84	9.68					
6	9.514	9.357	9.205	9.058	8.915	8.777	8.643	8.513	8.386	8.264					
7	8.144	8.028	7.916	7.806	7.700	7.596	7.495	7.396	7.300	7.207					
8	7.115	7.026	6.940	6.855	6.772	6.691	6.612	6.535	6.460	6.386					
9	6.314	6.243	6.174	6.107	6.041	5.976	5.912	5.850	5.789	5.730	1'	2'	3'	4'	5'
10	5.6713	5.6140	5.5578	5.5026	5.4486	5.3955	5.3435	5.2924	5.2422	5.1929	87	175	263	350	438
11	5.1446	5.0970	5.0504	5.0045	4.9594	4.9152	4.8716	4.8288	4.7867	4.7453	73	146	220	293	366
12	4.7046	4.6646	4.6252	4.5864	4.5483	4.5107	4.4737	4.4373	4.4015	4.3662	62	124	186	248	310
13	4.3315	4.2972	4.2635	4.2303	4.1976	4.1653	4.1335	4.1022	4.0713	4.0408	53	107	160	214	267
14	4.0108	3.9812	3.9520	3.9232	3.8947	3.8667	3.8391	3.8118	3.7848	3.7583	46	93	139	186	232
15	3.7321	3.7062	3.6806	3.6554	3.6305	3.6059	3.5816	3.5576	3.5339	3.5105	41	81	122	163	203
16	3.4874	3.4646	3.4420	3.4197	3.3977	3.3759	3.3544	3.3332	3.3122	3.2914	36	72	108	144	180
17	3.2709	3.2506	3.2305	3.2106	3.1910	3.1716	3.1524	3.1334	3.1146	3.0961	32	64	97	129	161
18	3.0777	3.0595	3.0415	3.0237	3.0061	2.9887	2.9714	2.9544	2.9375	2.9208	29	58	87	116	144
19	2.9042	2.8878	2.8716	2.8556	2.8397	2.8239	2.8083	2.7929	2.7776	2.7625	26	52	78	104	130
20	2.7475	2.7326	2.7179	2.7034	2.6889	2.6746	2.6605	2.6464	2.6325	2.6187	24	47	71	95	119
21	2.6051	2.5916	2.5782	2.5649	2.5517	2.5386	2.5257	2.5129	2.5002	2.4876	22	43	65	87	108
22	2.4751	2.4627	2.4504	2.4383	2.4262	2.4142	2.4023	2.3906	2.3789	2.3673	20	40	60	79	99
23	2.3559	2.3445	2.3332	2.3220	2.3109	2.2998	2.2889	2.2781	2.2673	2.2566	18	37	55	73	91
24	2.2460	2.2355	2.2251	2.2148	2.2045	2.1943	2.1842	2.1742	2.1642	2.1543	17	34	51	68	85
25	2.1445	2.1348	2.1251	2.1155	2.1060	2.0965	2.0872	2.0778	2.0686	2.0594	16	31	47	63	78
26	2.0503	2.0413	2.0323	2.0233	2.0145	2.0057	1.9970	1.9883	1.9797	1.9711	15	29	44	58	73
27	1.9626	1.9542	1.9458	1.9375	1.9292	1.9210	1.9128	1.9047	1.8967	1.8887	14	27	41	55	68
28	1.8807	1.8728	1.8650	1.8572	1.8495	1.8418	1.8341	1.8265	1.8190	1.8115	13	26	38	51	64
29	1.8040	1.7966	1.7893	1.7820	1.7747	1.7675	1.7603	1.7532	1.7461	1.7391	12	24	36	48	60
30	1.7321	1.7251	1.7182	1.7113	1.7045	1.6977	1.6909	1.6842	1.6775	1.6709	11	23	34	45	56
31	1.6643	1.6577	1.6512	1.6447	1.6383	1.6319	1.6255	1.6191	1.6128	1.6066	11	21	32	43	53
32	1.6003	1.5941	1.5880	1.5818	1.5757	1.5697	1.5637	1.5577	1.5517	1.5458	10	20	30	40	50
33	1.5399	1.5340	1.5282	1.5224	1.5166	1.5108	1.5051	1.4994	1.4938	1.4882	10	19	29	38	48
34	1.4826	1.4770	1.4715	1.4659	1.4605	1.4550	1.4496	1.4442	1.4388	1.4335	9	18	27	36	45
35	1.4281	1.4229	1.4176	1.4124	1.4071	1.4019	1.3968	1.3916	1.3865	1.3814	9	17	26	34	43
36	1.3764	1.3713	1.3663	1.3613	1.3564	1.3514	1.3465	1.3416	1.3367	1.3319	8	16	25	33	41
37	1.3270	1.3222	1.3175	1.3127	1.3079	1.3032	1.2985	1.2938	1.2892	1.2846	8	16	24	31	39
38	1.2799	1.2753	1.2708	1.2662	1.2617	1.2572	1.2527	1.2482	1.2437	1.2393	8	15	23	30	38
39	1.2349	1.2305	1.2261	1.2218	1.2174	1.2131	1.2088	1.2045	1.2002	1.1960	7	14	22	29	36
40	1.1918	1.1875	1.1833	1.1792	1.1750	1.1708	1.1667	1.1626	1.1585	1.1544	7	14	21	28	34
41	1.1504	1.1463	1.1423	1.1383	1.1343	1.1303	1.1263	1.1224	1.1184	1.1145	7	13	20	27	33
42	1.1106	1.1067	1.1028	1.0990	1.0951	1.0913	1.0875	1.0837	1.0799	1.0761	6	13	19	25	32
43	1.0724	1.0686	1.0649	1.0612	1.0575	1.0538	1.0501	1.0464	1.0428	1.0392	6	12	18	25	31
44	1.0355	1.0319	1.0283	1.0247	1.0212	1.0176	1.0141	1.0105	1.0070	1.0035	6	12	18	24	30

$$\cot A = \frac{1}{\tan A}$$

To find $\dfrac{1}{\tan 30° 22'}$ look up $\cot 30° 22' = 1.7068$

Natural Cotangents

Numbers in difference columns to be *subtracted*, not added.

°	0' 0.0°	6' 0.1°	12' 0.2°	18' 0.3°	24' 0.4°	30' 0.5°	36' 0.6°	42' 0.7°	48' 0.8°	54' 0.9°	1'	2'	3'	4'	5'
45	1.0000	0.9965	0.9930	0.9896	0.9861	0.9827	0.9793	0.9759	0.9725	0.9691	6	11	17	23	29
46	0.9657	0.9623	0.9590	0.9556	0.9523	0.9490	0.9457	0.9424	0.9391	0.9358	6	11	17	22	28
47	0.9325	0.9293	0.9260	0.9228	0.9195	0.9163	0.9131	0.9099	0.9067	0.9036	5	11	16	21	27
48	0.9004	0.8972	0.8941	0.8910	0.8878	0.8847	0.8816	0.8785	0.8754	0.8724	5	10	16	21	26
49	0.8693	0.8662	0.8632	0.8601	0.8571	0.8541	0.8511	0.8481	0.8451	0.8421	5	10	15	20	25
50	0.8391	0.8361	0.8332	0.8302	0.8273	0.8243	0.8214	0.8185	0.8156	0.8127	5	10	15	20	24
51	0.8098	0.8069	0.8040	0.8012	0.7983	0.7954	0.7926	0.7898	0.7869	0.7841	5	9	14	19	24
52	0.7813	0.7785	0.7757	0.7729	0.7701	0.7673	0.7646	0.7618	0.7590	0.7563	5	9	14	18	23
53	0.7536	0.7508	0.7481	0.7454	0.7427	0.7400	0.7373	0.7346	0.7319	0.7292	5	9	14	18	23
54	0.7265	0.7239	0.7212	0.7186	0.7159	0.7133	0.7107	0.7080	0.7054	0.7028	4	9	13	18	22
55	0.7002	0.6976	0.6950	0.6924	0.6899	0.6873	0.6847	0.6822	0.6796	0.6771	4	9	13	17	21
56	0.6745	0.6720	0.6694	0.6669	0.6644	0.6619	0.6594	0.6569	0.6544	0.6519	4	8	13	17	21
57	0.6494	0.6469	0.6445	0.6420	0.6395	0.6371	0.6346	0.6322	0.6297	0.6273	4	8	12	16	20
58	0.6249	0.6224	0.6200	0.6176	0.6152	0.6128	0.6104	0.6080	0.6056	0.6032	4	8	12	16	20
59	0.6009	0.5985	0.5961	0.5938	0.5914	0.5890	0.5867	0.5844	0.5820	0.5797	4	8	12	16	20
60	0.5774	0.5750	0.5727	0.5704	0.5681	0.5658	0.5635	0.5612	0.5589	0.5566	4	8	12	15	19
61	0.5543	0.5520	0.5498	0.5475	0.5452	0.5430	0.5407	0.5384	0.5362	0.5340	4	8	11	15	19
62	0.5317	0.5295	0.5272	0.5250	0.5228	0.5206	0.5184	0.5161	0.5139	0.5117	4	7	11	15	18
63	0.5095	0.5073	0.5051	0.5029	0.5008	0.4986	0.4964	0.4942	0.4921	0.4899	4	7	11	15	18
64	0.4877	0.4856	0.4834	0.4813	0.4791	0.4770	0.4748	0.4727	0.4706	0.4684	4	7	11	14	18
65	0.4663	0.4642	0.4621	0.4599	0.4578	0.4557	0.4536	0.4515	0.4494	0.4473	4	7	11	14	18
66	0.4452	0.4431	0.4411	0.4390	0.4369	0.4348	0.4327	0.4307	0.4286	0.4265	3	7	10	14	17
67	0.4245	0.4224	0.4204	0.4183	0.4163	0.4142	0.4122	0.4101	0.4081	0.4061	3	7	10	14	17
68	0.4040	0.4020	0.4000	0.3979	0.3959	0.3939	0.3919	0.3899	0.3879	0.3859	3	7	10	13	17
69	0.3839	0.3819	0.3799	0.3779	0.3759	0.3739	0.3719	0.3699	0.3679	0.3659	3	7	10	13	17
70	0.3640	0.3620	0.3600	0.3581	0.3561	0.3541	0.3522	0.3502	0.3482	0.3463	3	7	10	13	16
71	0.3443	0.3424	0.3404	0.3385	0.3365	0.3346	0.3327	0.3307	0.3288	0.3269	3	6	10	13	16
72	0.3249	0.3230	0.3211	0.3191	0.3172	0.3153	0.3134	0.3115	0.3096	0.3076	3	6	10	13	16
73	0.3057	0.3038	0.3019	0.3000	0.2981	0.2962	0.2943	0.2924	0.2905	0.2886	3	6	9	13	16
74	0.2867	0.2849	0.2830	0.2811	0.2792	0.2773	0.2754	0.2736	0.2717	0.2698	3	6	9	13	16
75	0.2679	0.2661	0.2642	0.2623	0.2605	0.2586	0.2568	0.2549	0.2530	0.2512	3	6	9	12	16
76	0.2493	0.2475	0.2456	0.2438	0.2419	0.2401	0.2382	0.2364	0.2345	0.2327	3	6	9	12	15
77	0.2309	0.2290	0.2272	0.2254	0.2235	0.2217	0.2199	0.2180	0.2162	0.2144	3	6	9	12	15
78	0.2126	0.2107	0.2089	0.2071	0.2053	0.2035	0.2016	0.1998	0.1980	0.1962	3	6	9	12	15
79	0.1944	0.1926	0.1908	0.1890	0.1871	0.1853	0.1835	0.1817	0.1799	0.1781	3	6	9	12	15
80	0.1763	0.1745	0.1727	0.1709	0.1691	0.1673	0.1655	0.1638	0.1620	0.1602	3	6	9	12	15
81	0.1584	0.1566	0.1548	0.1530	0.1512	0.1495	0.1477	0.1459	0.1441	0.1423	3	6	9	12	15
82	0.1405	0.1388	0.1370	0.1352	0.1334	0.1317	0.1299	0.1281	0.1263	0.1246	3	6	9	12	15
83	0.1228	0.1210	0.1192	0.1175	0.1157	0.1139	0.1122	0.1104	0.1086	0.1069	3	6	9	12	15
84	0.1051	0.1033	0.1016	0.0998	0.0981	0.0963	0.0945	0.0928	0.0910	0.0892	3	6	9	12	15
85	0.0875	0.0857	0.0840	0.0822	0.0805	0.0787	0.0769	0.0752	0.0734	0.0717	3	6	9	12	15
86	0.0699	0.0682	0.0664	0.0647	0.0629	0.0612	0.0594	0.0577	0.0559	0.0542	3	6	9	12	15
87	0.0524	0.0507	0.0489	0.0472	0.0454	0.0437	0.0419	0.0402	0.0384	0.0367	3	6	9	12	15
88	0.0349	0.0332	0.0314	0.0297	0.0279	0.0262	0.0244	0.0227	0.0209	0.0192	3	6	9	12	15
89	0.0175	0.0157	0.0140	0.0122	0.0105	0.0087	0.0070	0.0052	0.0035	0.0017	3	6	9	12	15

Quadrant	Angle	cot A =	Examples
first	0°–90°	cot A	cot 56° 17' = 0.6673
second	90°–180°	−cot(180° − A)	cot 123° 43' = −cot(180° − 123° 43')
third	180°–270°	cot(A − 180°)	= −cot 56° 17' = −0.6673
fourth	270°–360°	−tan(360° − A)	cot 236° 17' = cot(236° 17' − 180°)
			= cot 56° 17' = 0.6673
			cot 326° 34' = −cot(360° − 326° 34')
			= −cot 56° 17' = −0.6673.

Natural Secants

°	0' 0.0°	6' 0.1°	12' 0.2°	18' 0.3°	24' 0.4°	30' 0.5°	36' 0.6°	42' 0.7°	48' 0.8°	54' 0.9°	1'	2'	3'	4'	5'
0	1.0000	1.0000	1.0000	1.0000	1.0000	1.0000	1.0001	1.0001	1.0001	1.0001					
1	1.0002	1.0002	1.0002	1.0003	1.0003	1.0003	1.0004	1.0004	1.0005	1.0006					
2	1.0006	1.0007	1.0007	1.0008	1.0009	1.0010	1.0010	1.0011	1.0012	1.0013					
3	1.0014	1.0015	1.0016	1.0017	1.0018	1.0019	1.0020	1.0021	1.0022	1.0023	0	0	0	1	1
4	1.0024	1.0026	1.0027	1.0028	1.0030	1.0031	1.0032	1.0034	1.0035	1.0037	0	0	1	1	1
5	1.0038	1.0040	1.0041	1.0043	1.0045	1.0046	1.0048	1.0050	1.0051	1.0053	0	1	1	1	1
6	1.0055	1.0057	1.0059	1.0061	1.0063	1.0065	1.0067	1.0069	1.0071	1.0073	0	1	1	1	2
7	1.0075	1.0077	1.0079	1.0082	1.0084	1.0086	1.0089	1.0091	1.0093	1.0096	0	1	1	2	2
8	1.0098	1.0101	1.0103	1.0106	1.0108	1.0111	1.0114	1.0116	1.0119	1.0122	0	1	1	2	2
9	1.0125	1.0127	1.0130	1.0133	1.0136	1.0139	1.0142	1.0145	1.0148	1.0151	0	1	1	2	2
10	1.0154	1.0157	1.0161	1.0164	1.0167	1.0170	1.0174	1.0177	1.0180	0.0184	1	1	2	2	3
11	1.0187	1.0191	1.0194	1.0198	1.0201	1.0205	1.0209	1.0212	1.0216	1.0220	1	1	2	2	3
12	1.0223	1.0227	1.0231	1.0235	1.0239	1.0243	1.0247	1.0251	1.0255	1.0259	1	1	2	3	3
13	1.0263	1.0267	1.0271	1.0276	1.0280	1.0284	1.0288	1.0293	1.0297	1.0302	1	1	2	3	4
14	1.0306	1.0311	1.0315	1.0320	1.0324	1.0329	1.0334	1.0338	1.0343	1.0348	1	2	2	3	4
15	1.0353	1.0358	1.0363	1.0367	1.0372	1.0377	1.0382	1.0388	1.0393	1.0398	1	2	3	3	4
16	1.0403	1.0408	1.0413	1.0419	1.0424	1.0429	1.0435	1.0440	1.0446	1.0451	1	2	3	4	4
17	1.0457	1.0463	1.0468	1.0474	1.0480	1.0485	1.0491	1.0497	1.0503	1.0509	1	2	3	4	5
18	1.0515	1.0521	1.0527	1.0533	1.0539	1.0545	1.0551	1.0557	1.0564	1.0570	1	2	3	4	5
19	1.0576	1.0583	1.0589	1.0595	1.0602	1.0608	1.0615	1.0622	1.0628	1.0635	1	2	3	4	5
20	1.0642	1.0649	1.0655	1.0662	1.0669	1.0676	1.0683	1.0690	1.0697	1.0704	1	2	3	5	6
21	1.0711	1.0719	1.0726	1.0733	1.0740	1.0748	1.0755	1.0763	1.0770	1.0778	1	2	4	5	6
22	1.0785	1.0793	1.0801	1.0808	1.0816	1.0824	1.0832	1.0840	1.0848	1.0856	1	3	4	5	7
23	1.0864	1.0872	1.0880	1.0888	1.0896	1.0904	1.0913	1.0921	1.0929	1.0938	1	3	4	6	7
24	1.0946	1.0955	1.0963	1.0972	1.0981	1.0989	1.0998	1.1007	1.1016	1.1025	1	3	4	6	7
25	1.1034	1.1043	1.1052	1.1061	1.1070	1.1079	1.1089	1.1098	1.1107	1.1117	2	3	5	6	8
26	1.1126	1.1136	1.1145	1.1155	1.1164	1.1174	1.1184	1.1194	1.1203	1.1213	2	3	5	6	8
27	1.1223	1.1233	1.1243	1.1253	1.1264	1.1274	1.1284	1.1294	1.1305	1.1315	2	3	5	7	9
28	1.1326	1.1336	1.1347	1.1357	1.1368	1.1379	1.1390	1.1401	1.1412	1.1423	2	4	5	7	9
29	1.1434	1.1445	1.1456	1.1467	1.1478	1.1490	1.1501	1.1512	1.1524	1.1535	2	4	6	8	9
30	1.1547	1.1559	1.1570	1.1582	1.1594	1.1606	1.1618	1.1630	1.1642	1.1654	2	4	6	8	10
31	1.1666	1.1679	1.1691	1.1703	1.1716	1.1728	1.1741	1.1753	1.1766	1.1779	2	4	6	8	10
32	1.1792	1.1805	1.1818	1.1831	1.1844	1.1857	1.1870	1.1883	1.1897	1.1910	2	4	7	9	11
33	1.1924	1.1937	1.1951	1.1964	1.1978	1.1992	1.2006	1.2020	1.2034	1.2048	2	5	7	9	12
34	1.2062	1.2076	1.2091	1.2105	1.2120	1.2134	1.2149	1.2163	1.2178	1.2193	2	5	7	10	12
35	1.2208	1.2223	1.2238	1.2253	1.2268	1.2283	1.2299	1.2314	1.2329	1.2345	3	5	8	10	13
36	1.2361	1.2376	1.2392	1.2408	1.2424	1.2440	1.2456	1.2472	1.2489	1.2505	3	5	8	11	13
37	1.2521	1.2538	1.2554	1.2571	1.2588	1.2605	1.2622	1.2639	1.2656	1.2673	3	6	8	11	14
38	1.2690	1.2708	1.2725	1.2742	1.2760	1.2778	1.2796	1.2813	1.2831	1.2849	3	6	9	12	15
39	1.2868	1.2886	1.2904	1.2923	1.2941	1.2960	1.2978	1.2997	1.3016	1.3035	3	6	9	12	16
40	1.3054	1.3073	1.3093	1.3112	1.3131	1.3151	1.3171	1.3190	1.3210	1.3230	3	7	10	13	16
41	1.3250	1.3270	1.3291	1.3311	1.3331	1.3352	1.3373	1.3393	1.3414	1.3435	3	7	10	14	17
42	1.3456	1.3478	1.3499	1.3520	1.3542	1.3563	1.3585	1.3607	1.3629	1.3651	4	7	11	14	18
43	1.3673	1.3696	1.3718	1.3741	1.3763	1.3786	1.3809	1.3832	1.3855	1.3878	4	8	11	15	19
44	1.3902	1.3925	1.3949	1.3972	1.3996	1.4020	1.4044	1.4069	1.4093	1.4118	4	8	12	16	20

$$\sec A = \frac{1}{\cos A}$$

To find $\dfrac{1}{\cos 35° 40'}$ look up $\sec 35° 40' = 1.2309$

Natural Secants

°	0' 0.0°	6' 0.1°	12' 0.2°	18' 0.3°	24' 0.4°	30' 0.5°	36' 0.6°	42' 0.7°	48' 0.8°	54' 0.9°	1'	2'	3'	4'	5'
45	1.4142	1.4167	1.4192	1.4217	1.4242	1.4267	1.4293	1.4318	1.4344	1.4370	4	8	13	17	21
46	1.4396	1.4422	1.4448	1.4474	1.4501	1.4527	1.4554	1.4581	1.4608	1.4635	4	9	13	18	22
47	1.4663	1.4690	1.4718	1.4746	1.4774	1.4802	1.4830	1.4859	1.4887	1.4916	5	9	14	19	23
48	1.4945	1.4974	1.5003	1.5032	1.5062	1.5092	1.5121	1.5151	1.5182	1.5212	5	10	15	20	25
49	1.5243	1.5273	1.5304	1.5335	1.5366	1.5398	1.5429	1.5461	1.5493	1.5525	5	10	16	21	26
50	1.5557	1.5590	1.5622	1.5655	1.5688	1.5721	1.5755	1.5788	1.5822	1.5856	6	11	17	22	28
51	1.5890	1.5925	1.5959	1.5994	1.6029	1.6064	1.6099	1.6135	1.6171	1.6207	6	12	18	23	29
52	1.6243	1.6279	1.6316	1.6353	1.6390	1.6427	1.6464	1.6502	1.6540	1.6578	6	12	19	25	31
53	1.6616	1.6655	1.6694	1.6733	1.6772	1.6812	1.6852	1.6892	1.6932	1.6972	7	13	20	26	33
54	1.7013	1.7054	1.7095	1.7137	1.7179	1.7221	1.7263	1.7305	1.7348	1.7391	7	14	21	28	35
55	1.7434	1.7478	1.7522	1.7566	1.7610	1.7655	1.7700	1.7745	1.7791	1.7837	7	15	22	30	37
56	1.7883	1.7929	1.7976	1.8023	1.8070	1.8118	1.8166	1.8214	1.8263	1.8312	8	16	24	32	40
57	1.8361	1.8410	1.8460	1.8510	1.8561	1.8612	1.8663	1.8714	1.8766	1.8818	8	17	25	34	42
58	1.8871	1.8924	1.8977	1.9031	1.9084	1.9139	1.9194	1.9249	1.9304	1.9360	9	18	27	36	45
59	1.9416	1.9473	1.9530	1.9587	1.9645	1.9703	1.9762	1.9821	1.9880	1.9940	10	19	29	39	49
60	2.0000	2.0061	2.0122	2.0183	2.0245	2.0308	2.0371	2.0434	2.0498	2.0562	10	21	31	42	52
61	2.0627	2.0692	2.0757	2.0824	2.0890	2.0957	2.1025	2.1093	2.1162	2.1231	11	22	34	45	56
62	2.1301	2.1371	2.1441	2.1513	2.1584	2.1657	2.1730	2.1803	2.1877	2.1952	12	24	36	48	61
63	2.2027	2.2103	2.2179	2.2256	2.2333	2.2412	2.2490	2.2570	2.2650	2.2730	13	26	39	52	65
64	2.2812	2.2894	2.2976	2.3060	2.3144	2.3228	2.3314	2.3400	2.3486	2.3574	14	28	43	57	71
65	2.3662	2.3751	2.3841	2.3931	2.4022	2.4114	2.4207	2.4300	2.4395	2.4490	15	31	46	62	77
66	2.4586	2.4683	2.4780	2.4879	2.4978	2.5078	2.5180	2.5282	2.5384	2.5488	17	34	50	67	84
67	2.5593	2.5699	2.5805	2.5913	2.6022	2.6131	2.6242	2.6354	2.6466	2.6580	18	37	55	73	92
68	2.6695	2.6811	2.6927	2.7046	2.7165	2.7285	2.7407	2.7529	2.7653	2.7778	20	40	60	81	101
69	2.7904	2.8032	2.8161	2.8291	2.8422	2.8555	2.8688	2.8824	2.8960	2.9099	22	44	67	89	111
70	2.9238	2.9379	2.9521	2.9665	2.9811	2.9957	3.0106	3.0256	3.0407	3.0561	25	49	74	98	123
71	3.0716	3.0872	3.1030	3.1190	3.1352	3.1515	3.1681	3.1848	3.2017	3.2188	27	55	82	110	137
72	3.2361	3.2535	3.2712	3.2891	3.3072	3.3255	3.3440	3.3628	3.3817	3.4009	31	61	92	123	153
73	3.4203	3.4399	3.4598	3.4799	3.5003	3.5209	3.5418	3.5629	3.5843	3.6060	35	69	104	138	173
74	3.6280	3.6502	3.6727	3.6955	3.7186	3.7420	3.7657	3.7897	3.8140	3.8387	39	79	118	157	196
75	3.8637	3.8890	3.9147	3.9408	3.9672	3.9939	4.0211	4.0486	4.0765	4.1048	45	90	135	180	225
76	4.1336	4.1627	4.1923	4.2223	4.2527	4.2837	4.3150	4.3469	4.3792	4.4121	52	104	156	207	260
77	4.4454	4.4793	4.5137	4.5486	4.5841	4.6202	4.6569	4.6942	4.7321	4.7706	61	121	182	242	303
78	4.8097	4.8496	4.8901	4.9313	4.9732	5.0159	5.0593	5.1034	5.1484	5.1942	72	143	215	287	359
79	5.2408	5.2883	5.3367	5.3860	5.4362	5.4874	5.5396	5.5928	5.6470	5.7023	86	172	258	344	431
80	5.759	5.816	5.875	5.935	5.996	6.059	6.123	6.188	6.255	6.323					
81	6.392	6.464	6.537	6.611	6.687	6.765	6.845	6.927	7.011	7.097					
82	7.185	7.276	7.368	7.463	7.561	7.661	7.764	7.870	7.979	8.091					
83	8.206	8.324	8.446	8.571	8.700	8.834	8.971	9.113	9.259	9.411					
84	9.57	9.73	9.90	10.07	10.25	10.43	10.63	10.83	11.03	11.25		Differences			
85	11.47	11.71	11.95	12.20	12.47	12.75	13.03	13.34	13.65	13.99		untrustworthy			
86	14.34	14.70	15.09	15.50	15.93	16.38	16.86	17.37	17.91	18.49		here			
87	19.11	19.77	20.47	21.23	22.03	22.93	23.88	24.92	26.05	27.29					
88	28.65	30.16	31.84	33.71	35.81	38.20	40.93	44.08	47.75	52.09					
89	57.30	63.66	71.62	81.85	95.49	114.6	143.2	191.0	286.5	573.0					

Quadrant	Angle	sec A	Examples
first	0–90°	sec A	sec 33° 26' = 1.1983
second	90°–180°	−sec(180° − A)	sec 146° 34' = −sec(180° − 146° 34')
third	180°–270°	−sec(A − 180°)	= −sec 33° 26' = −1.1983
fourth	270°–360°	sec(360° − A)	sec 213° 26' = −sec(213° 26' − 180°)
			= −sec 33° 26' = −1.1983
			sec 326° 34' = sec(360° − 326° 34')
			= sec 33° 26' = 1.1983

63

Natural Cosecants

Numbers in difference columns to be *subtracted*, not added.

°	0' 0.0°	6' 0.1°	12' 0.2°	18' 0.3°	24' 0.4°	30' 0.5°	36' 0.6°	42' 0.7°	48' 0.8°	54' 0.9°	1'	2'	3'	4'	5'
0	∞	573.0	286.5	191.0	143.2	114.6	95.49	81.85	71.62	63.66					
1	57.30	52.09	47.75	44.08	40.93	38.20	35.81	33.71	31.84	30.16		Differences			
2	28.65	27.29	26.05	24.92	23.88	22.93	22.04	21.23	20.47	19.77		untrustworthy			
3	19.11	18.49	17.91	17.37	16.86	16.38	15.93	15.50	15.09	14.70		here			
4	14.34	13.99	13.65	13.34	13.03	12.75	12.47	12.20	11.95	11.71					
5	11.47	11.25	11.03	10.83	10.63	10.43	10.25	10.07	9.90	9.73					
6	9.567	9.411	9.259	9.113	8.971	8.834	8.700	8.571	8.446	8.324					
7	8.206	8.091	7.979	7.870	7.764	7.661	7.561	7.463	7.368	7.276					
8	7.185	7.097	7.011	6.927	6.845	6.765	6.687	6.611	6.537	6.464					
9	6.392	6.323	6.255	6.188	6.123	6.059	5.996	5.935	5.875	5.816					
10	5.7588	5.7023	5.6470	5.5928	5.5396	5.4874	5.4362	5.3860	5.3367	5.2883	86	172	258	344	431
11	5.2408	5.1942	5.1484	5.1034	5.0593	5.0159	4.9732	4.9313	4.8901	4.8496	72	143	215	287	359
12	4.8097	4.7706	4.7321	4.6942	4.6569	4.6202	4.5841	4.5486	4.5137	4.4793	61	121	182	242	303
13	4.4454	4.4121	4.3792	4.3469	4.3150	4.2837	4.2527	4.2223	4.1923	4.1627	52	104	156	207	260
14	4.1336	4.1048	4.0765	4.0486	4.0211	3.9939	3.9672	3.9408	3.9147	3.8890	45	90	135	180	225
15	3.8637	3.8387	3.8140	3.7897	3.7657	3.7420	3.7186	3.6955	3.6727	3.6502	39	79	118	157	196
16	3.6280	3.6060	3.5843	3.5629	3.5418	3.5209	3.5003	3.4799	3.4598	3.4399	35	69	104	138	173
17	3.4203	3.4009	3.3817	3.3628	3.3440	3.3255	3.3072	3.2891	3.2712	3.2535	31	61	92	123	153
18	3.2361	3.2188	3.2017	3.1848	3.1681	3.1515	3.1352	3.1190	3.1030	3.0872	27	55	82	110	137
19	3.0716	3.0561	3.0407	3.0256	3.0106	2.9957	2.9811	2.9665	2.9521	2.9379	25	49	74	98	123
20	2.9238	2.9099	2.8960	2.8824	2.8688	2.8555	2.8422	2.8291	2.8161	2.8032	22	44	67	89	111
21	2.7904	2.7778	2.7653	2.7529	2.7407	2.7285	2.7165	2.7046	2.6927	2.6811	20	40	60	81	101
22	2.6695	2.6580	2.6466	2.6354	2.6242	2.6131	2.6022	2.5913	2.5805	2.5699	18	37	55	73	92
23	2.5593	2.5488	2.5384	2.5282	2.5180	2.5078	2.4978	2.4879	2.4780	2.4683	17	34	50	67	84
24	2.4586	2.4490	2.4395	2.4300	2.4207	2.4114	2.4022	2.3931	2.3841	2.3751	15	31	46	62	77
25	2.3662	2.3574	2.3486	2.3400	2.3314	2.3228	2.3144	2.3060	2.2976	2.2894	14	28	43	57	71
26	2.2812	2.2730	2.2650	2.2570	2.2490	2.2412	2.2333	2.2256	2.2179	2.2103	13	26	39	52	65
27	2.2027	2.1952	2.1877	2.1803	2.1730	2.1657	2.1584	2.1513	2.1441	2.1371	12	24	36	48	61
28	2.1301	2.1231	2.1162	2.1093	2.1025	2.0957	2.0890	2.0824	2.0757	2.0692	11	22	34	45	56
29	2.0627	2.0562	2.0498	2.0434	2.0371	2.0308	2.0245	2.0183	2.0122	2.0061	10	21	31	42	52
30	2.0000	1.9940	1.9880	1.9821	1.9762	1.9703	1.9645	1.9587	1.9530	1.9473	10	19	29	39	49
31	1.9416	1.9360	1.9304	1.9249	1.9194	1.9139	1.9084	1.9031	1.8977	1.8924	9	18	27	36	45
32	1.8871	1.8818	1.8766	1.8714	1.8663	1.8612	1.8561	1.8510	1.8460	1.8410	8	17	25	34	42
33	1.8361	1.8312	1.8263	1.8214	1.8166	1.8118	1.8070	1.8023	1.7976	1.7929	8	16	24	32	40
34	1.7883	1.7837	1.7791	1.7745	1.7700	1.7655	1.7610	1.7566	1.7522	1.7478	7	15	22	30	37
35	1.7434	1.7391	1.7348	1.7305	1.7263	1.7221	1.7179	1.7137	1.7095	1.7054	7	14	21	28	35
36	1.7013	1.6972	1.6932	1.6892	1.6852	1.6812	1.6772	1.6733	1.6694	1.6655	7	13	20	26	33
37	1.6616	1.6578	1.6540	1.6502	1.6464	1.6427	1.6390	1.6353	1.6316	1.6279	6	12	19	25	31
38	1.6243	1.6207	1.6171	1.6135	1.6099	1.6064	1.6029	1.5994	1.5959	1.5925	6	12	18	23	29
39	1.5890	1.5856	1.5822	1.5788	1.5755	1.5721	1.5688	1.5655	1.5622	1.5590	6	11	17	22	28
40	1.5557	1.5525	1.5493	1.5461	1.5429	1.5398	1.5366	1.5335	1.5304	1.5273	5	10	16	21	26
41	1.5243	1.5212	1.5182	1.5151	1.5121	1.5092	1.5062	1.5032	1.5003	1.4974	5	10	15	20	25
42	1.4945	1.4916	1.4887	1.4859	1.4830	1.4802	1.4774	1.4746	1.4718	1.4690	5	9	14	19	23
43	1.4663	1.4635	1.4608	1.4581	1.4554	1.4527	1.4501	1.4474	1.4448	1.4422	4	9	13	18	22
44	1.4396	1.4370	1.4344	1.4318	1.4293	1.4267	1.4242	1.4217	1.4192	1.4167	4	8	13	17	21

$$\operatorname{cosec} A = \frac{1}{\sin A}$$

To find $\dfrac{1}{\sin 38° 23'}$ look up $\operatorname{cosec} 38° 23' = 1.6105$

Natural Cosecants

Numbers in difference columns to be *subtracted*, not added.

°	0' 0.0°	6' 0.1°	12' 0.2°	18' 0.3°	24' 0.4°	30' 0.5°	36' 0.6°	42' 0.7°	48' 0.8°	54' 0.9°	1'	2'	3'	4'	5'
45	1.4142	1.4118	1.4093	1.4069	1.4044	1.4020	1.3996	1.3972	1.3949	1.3925	4	8	12	16	20
46	1.3902	1.3878	1.3855	1.3832	1.3809	1.3786	1.3763	1.3741	1.3718	1.3696	4	8	11	15	19
47	1.3673	1.3651	1.3629	1.3607	1.3585	1.3563	1.3542	1.3520	1.3499	1.3478	4	7	11	14	18
48	1.3456	1.3435	1.3414	1.3393	1.3373	1.3352	1.3331	1.3311	1.3291	1.3270	3	7	10	14	17
49	1.3250	1.3230	1.3210	1.3190	1.3171	1.3151	1.3131	1.3112	1.3093	1.3073	3	7	10	13	17
50	1.3054	1.3035	1.3016	1.2997	1.2978	1.2960	1.2941	1.2923	1.2904	1.2886	3	6	9	12	16
51	1.2868	1.2849	1.2831	1.2813	1.2796	1.2778	1.2760	1.2742	1.2725	1.2708	3	6	9	12	15
52	1.2690	1.2673	1.2656	1.2639	1.2622	1.2605	1.2588	1.2571	1.2554	1.2538	3	6	8	11	14
53	1.2521	1.2505	1.2489	1.2472	1.2456	1.2440	1.2424	1.2408	1.2392	1.2376	3	5	8	11	13
54	1.2361	1.2345	1.2329	1.2314	1.2299	1.2283	1.2268	1.2253	1.2238	1.2223	3	5	8	10	13
55	1.2208	1.2193	1.2178	1.2163	1.2149	1.2134	1.2120	1.2105	1.2091	1.2076	2	5	7	10	12
56	1.2062	1.2048	1.2034	1.2020	1.2006	1.1992	1.1978	1.1964	1.1951	1.1937	2	5	7	9	12
57	1.1924	1.1910	1.1897	1.1883	1.1870	1.1857	1.1844	1.1831	1.1818	1.1805	2	4	7	9	11
58	1.1792	1.1779	1.1766	1.1753	1.1741	1.1728	1.1716	1.1703	1.1691	1.1679	2	4	6	8	10
59	1.1666	1.1654	1.1642	1.1630	1.1618	1.1606	1.1594	1.1582	1.1570	1.1559	2	4	6	8	10
60	1.1547	1.1535	1.1524	1.1512	1.1501	1.1490	1.1478	1.1467	1.1456	1.1445	2	4	6	7	9
61	1.1434	1.1423	1.1412	1.1401	1.1390	1.1379	1.1368	1.1357	1.1347	1.1336	2	4	5	7	9
62	1.1326	1.1315	1.1305	1.1294	1.1284	1.1274	1.1264	1.1253	1.1243	1.1233	2	3	5	7	9
63	1.1223	1.1213	1.1203	1.1194	1.1184	1.1174	1.1164	1.1155	1.1145	1.1136	2	3	5	6	8
64	1.1126	1.1117	1.1107	1.1098	1.1089	1.1079	1.1070	1.1061	1.1052	1.1043	2	3	5	6	8
65	1.1034	1.1025	1.1016	1.1007	1.0998	1.0989	1.0981	1.0972	1.0963	1.0955	1	3	4	6	7
66	1.0946	1.0938	1.0929	1.0921	1.0913	1.0904	1.0896	1.0888	1.0880	1.0872	1	3	4	5	7
67	1.0864	1.0856	1.0848	1.0840	1.0832	1.0824	1.0816	1.0808	1.0801	1.0793	1	3	4	5	7
68	1.0785	1.0778	1.0770	1.0763	1.0755	1.0748	1.0740	1.0733	1.0726	1.0719	1	2	4	5	6
69	1.0711	1.0704	1.0697	1.0690	1.0683	1.0676	1.0669	1.0662	1.0655	1.0649	1	2	3	5	6
70	1.0642	1.0635	1.0628	1.0622	1.0615	1.0608	1.0602	1.0595	1.0589	1.0583	1	2	3	4	5
71	1.0576	1.0570	1.0564	1.0557	1.0551	1.0545	1.0539	1.0533	1.0527	1.0521	1	2	3	4	5
72	1.0515	1.0509	1.0503	1.0497	1.0491	1.0485	1.0480	1.0474	1.0468	1.0463	1	2	3	4	5
73	1.0457	1.0451	1.0446	1.0440	1.0435	1.0429	1.0424	1.0419	1.0413	1.0408	1	2	3	4	4
74	1.0403	1.0398	1.0393	1.0388	1.0382	1.0377	1.0372	1.0367	1.0363	1.0358	1	2	2	3	4
75	1.0353	1.0348	1.0343	1.0338	1.0334	1.0329	1.0324	1.0320	1.0315	1.0311	1	2	2	3	4
76	1.0306	1.0302	1.0297	1.0293	1.0288	1.0284	1.0280	1.0276	1.0271	1.0267	1	1	2	3	4
77	1.0263	1.0259	1.0255	1.0251	1.0247	1.0243	1.0239	1.0235	1.0231	1.0227	1	1	2	3	3
78	1.0223	1.0220	1.0216	1.0212	1.0209	1.0205	1.0201	1.0198	1.0194	1.0191	1	1	2	2	3
79	1.0187	1.0184	1.0180	1.0177	1.0174	1.0170	1.0167	1.0164	1.0161	1.0157	1	1	2	2	3
80	1.0154	1.0151	1.0148	1.0145	1.0142	1.0139	1.0136	1.0133	1.0130	1.0127	0	1	1	2	2
81	1.0125	1.0122	1.0119	1.0116	1.0114	1.0111	1.0108	1.0106	1.0103	1.0101	0	1	1	2	2
82	1.0098	1.0096	1.0093	1.0091	1.0089	1.0086	1.0084	1.0082	1.0079	1.0077	0	1	1	2	2
83	1.0075	1.0073	1.0071	1.0069	1.0067	1.0065	1.0063	1.0061	1.0059	1.0057	0	1	1	1	2
84	1.0055	1.0053	1.0051	1.0050	1.0048	1.0046	1.0045	1.0043	1.0041	1.0040	0	1	1	1	1
85	1.0038	1.0037	1.0035	1.0034	1.0032	1.0031	1.0030	1.0028	1.0027	1.0026	0	0	1	1	1
86	1.0024	1.0023	1.0022	1.0021	1.0020	1.0019	1.0018	1.0017	1.0016	1.0015	0	0	0	0	1
87	1.0014	1.0013	1.0012	1.0011	1.0010	1.0010	1.0009	1.0008	1.0007	1.0007					
88	1.0006	1.0006	1.0005	1.0004	1.0004	1.0003	1.0003	1.0003	1.0002	1.0002		Differences			
89	1.0002	1.0001	1.0001	1.0001	1.0001	1.0000	1.0000	1.0000	1.0000	1.0000		untrustworthy			
												here			

Quadrant	Angle	cosec A =	Examples
first	0°–90°	cosec A	cosec 34° 38′ = 1.7595
second	90°–180°	cosec(180° − A)	cosec 145° 22′ = cosec(180° − 145° 22′)
third	180°–270°	− cosec(A − 180°)	= cosec 34° 38′ = 1.7595
fourth	270°–360°	− cosec(360° − A)	cosec 214° 38′ = − cosec(214° 38′ − 180°)
			= − cosec 34° 38′ = − 1.7595
			cosec 325° 22′ = − cosec(360° − 325° 22′)
			= − cosec 34° 38′ = − 1.7595

Degrees to Radians

°	0' 0.0°	6' 0.1°	12' 0.2°	18' 0.3°	24' 0.4°	30' 0.5°	36' 0.6°	42' 0.7°	48' 0.8°	54' 0.9°	1'	2'	3'	4'	5'
0	0.0000	0.0017	0.0035	0.0052	0.0070	0.0087	0.0105	0.0122	0.0140	0.0157	3	6	9	12	15
1	0.0175	0.0192	0.0209	0.0227	0.0244	0.0262	0.0279	0.0297	0.0314	0.0332	3	6	9	12	15
2	0.0349	0.0367	0.0384	0.0401	0.0419	0.0436	0.0454	0.0471	0.0489	0.0506	3	6	9	12	15
3	0.0524	0.0541	0.0559	0.0576	0.0593	0.0611	0.0628	0.0646	0.0663	0.0681	3	6	9	12	15
4	0.0698	0.0716	0.0733	0.0750	0.0768	0.0785	0.0803	0.0820	0.0838	0.0855	3	6	9	12	15
5	0.0873	0.0890	0.0908	0.0925	0.0942	0.0960	0.0977	0.0995	0.1012	0.1030	3	6	9	12	15
6	0.1047	0.1065	0.1082	0.1100	0.1117	0.1134	0.1152	0.1169	0.1187	0.1204	3	6	9	12	15
7	0.1222	0.1239	0.1257	0.1274	0.1292	0.1309	0.1326	0.1344	0.1361	0.1379	3	6	9	12	15
8	0.1396	0.1414	0.1431	0.1449	0.1466	0.1484	0.1501	0.1518	0.1536	0.1553	3	6	9	12	15
9	0.1571	0.1588	0.1606	0.1623	0.1641	0.1658	0.1676	0.1693	0.1710	0.1728	3	6	9	12	15
10	0.1745	0.1763	0.1780	0.1798	0.1815	0.1833	0.1850	0.1868	0.1885	0.1902	3	6	9	12	15
11	0.1920	0.1937	0.1955	0.1972	0.1990	0.2007	0.2025	0.2042	0.2060	0.2077	3	6	9	12	15
12	0.2094	0.2112	0.2129	0.2147	0.2164	0.2182	0.2199	0.2217	0.2234	0.2251	3	6	9	12	15
13	0.2269	0.2286	0.2304	0.2321	0.2339	0.2356	0.2374	0.2391	0.2409	0.2426	3	6	9	12	15
14	0.2443	0.2461	0.2478	0.2496	0.2513	0.2531	0.2548	0.2566	0.2583	0.2601	3	6	9	12	15
15	0.2618	0.2635	0.2653	0.2670	0.2688	0.2705	0.2723	0.2740	0.2758	0.2775	3	6	9	12	15
16	0.2793	0.2810	0.2827	0.2845	0.2862	0.2880	0.2897	0.2915	0.2932	0.2950	3	6	9	12	15
17	0.2967	0.2985	0.3002	0.3019	0.3037	0.3054	0.3072	0.3089	0.3107	0.3124	3	6	9	12	15
18	0.3142	0.3159	0.3176	0.3194	0.3211	0.3229	0.3246	0.3264	0.3281	0.3299	3	6	9	12	15
19	0.3316	0.3334	0.3351	0.3368	0.3386	0.3403	0.3421	0.3438	0.3456	0.3473	3	6	9	12	15
20	0.3491	0.3508	0.3526	0.3543	0.3560	0.3578	0.3595	0.3613	0.3630	0.3648	3	6	9	12	15
21	0.3665	0.3683	0.3700	0.3718	0.3735	0.3752	0.3770	0.3787	0.3805	0.3822	3	6	9	12	15
22	0.3840	0.3857	0.3875	0.3892	0.3910	0.3927	0.3944	0.3962	0.3979	0.3997	3	6	9	12	15
23	0.4014	0.4032	0.4049	0.4067	0.4084	0.4102	0.4119	0.4136	0.4154	0.4171	3	6	9	12	15
24	0.4189	0.4206	0.4224	0.4241	0.4259	0.4276	0.4294	0.4311	0.4328	0.4346	3	6	9	12	15
25	0.4363	0.4381	0.4398	0.4416	0.4433	0.4451	0.4468	0.4485	0.4503	0.4520	3	6	9	12	15
26	0.4538	0.4555	0.4573	0.4590	0.4608	0.4625	0.4643	0.4660	0.4677	0.4695	3	6	9	12	15
27	0.4712	0.4730	0.4747	0.4765	0.4782	0.4800	0.4817	0.4835	0.4852	0.4869	3	6	9	12	15
28	0.4887	0.4904	0.4922	0.4939	0.4957	0.4974	0.4992	0.5009	0.5027	0.5044	3	6	9	12	15
29	0.5061	0.5079	0.5096	0.5114	0.5131	0.5149	0.5166	0.5184	0.5201	0.5219	3	6	9	12	15
30	0.5236	0.5253	0.5271	0.5288	0.5306	0.5323	0.5341	0.5358	0.5376	0.5393	3	6	9	12	15
31	0.5411	0.5428	0.5445	0.5463	0.5480	0.5498	0.5515	0.5533	0.5550	0.5568	3	6	9	12	15
32	0.5585	0.5603	0.5620	0.5637	0.5655	0.5672	0.5690	0.5707	0.5725	0.5742	3	6	9	12	15
33	0.5760	0.5777	0.5794	0.5812	0.5829	0.5847	0.5864	0.5882	0.5899	0.5917	3	6	9	12	15
34	0.5934	0.5952	0.5969	0.5986	0.6004	0.6021	0.6039	0.6056	0.6074	0.6091	3	6	9	12	15
35	0.6109	0.6126	0.6144	0.6161	0.6178	0.6196	0.6213	0.6231	0.6248	0.6266	3	6	9	12	15
36	0.6283	0.6301	0.6318	0.6336	0.6353	0.6370	0.6388	0.6405	0.6423	0.6440	3	6	9	12	15
37	0.6458	0.6475	0.6493	0.6510	0.6528	0.6545	0.6562	0.6580	0.6597	0.6615	3	6	9	12	15
38	0.6632	0.6650	0.6667	0.6685	0.6702	0.6720	0.6737	0.6754	0.6772	0.6789	3	6	9	12	15
39	0.6807	0.6824	0.6842	0.6859	0.6877	0.6894	0.6912	0.6929	0.6946	0.6964	3	6	9	12	15
40	0.6981	0.6999	0.7016	0.7034	0.7051	0.7069	0.7086	0.7103	0.7121	0.7138	3	6	9	12	15
41	0.7156	0.7173	0.7191	0.7208	0.7226	0.7243	0.7261	0.7278	0.7295	0.7313	3	6	9	12	15
42	0.7330	0.7348	0.7365	0.7383	0.7400	0.7418	0.7435	0.7453	0.7470	0.7487	3	6	9	12	15
43	0.7505	0.7522	0.7540	0.7557	0.7575	0.7592	0.7610	0.7627	0.7645	0.7662	3	6	9	12	15
44	0.7679	0.7697	0.7714	0.7732	0.7749	0.7767	0.7784	0.7802	0.7819	0.7837	3	6	9	12	15
45	0.7854	0.7871	0.7889	0.7906	0.7924	0.7941	0.7959	0.7976	0.7994	0.8011	3	6	9	12	15

$$\theta° = \frac{\pi \times \theta°}{180} = 0.01745 \times \theta° \text{ radians}$$

$$30° = \frac{\pi}{6} \text{radians} \qquad 60° = \frac{\pi}{3} \text{radians}$$

$$45° = \frac{\pi}{4} \text{ radians} \qquad 90° = \frac{\pi}{2} \text{radians}$$

Degrees to Radians

°	0' 0.0°	6' 0.1°	12' 0.2°	18' 0.3°	24' 0.4°	30' 0.5°	36' 0.6°	42' 0.7°	48' 0.8°	54' 0.9°	1'	2'	3'	4'	5'
45	0.7854	0.7871	0.7889	0.7906	0.7924	0.7941	0.7959	0.7976	0.7994	0.8011	3	6	9	12	15
46	0.8029	0.8046	0.8063	0.8081	0.8098	0.8116	0.8133	0.8151	0.8168	0.8186	3	6	9	12	15
47	0.8203	0.8221	0.8238	0.8255	0.8273	0.8290	0.8308	0.8325	0.8343	0.8360	3	6	9	12	15
48	0.8378	0.8395	0.8412	0.8430	0.8447	0.8465	0.8482	0.8500	0.8517	0.8535	3	6	9	12	15
49	0.8552	0.8570	0.8587	0.8604	0.8622	0.8639	0.8657	0.8674	0.8692	0.8709	3	6	9	12	15
50	0.8727	0.8744	0.8762	0.8779	0.8796	0.8814	0.8831	0.8849	0.8866	0.8884	3	6	9	12	15
51	0.8901	0.8919	0.8936	0.8954	0.8971	0.8988	0.9006	0.9023	0.9041	0.9058	3	6	9	12	15
52	0.9076	0.9093	0.9111	0.9128	0.9146	0.9163	0.9180	0.9198	0.9215	0.9233	3	6	9	12	15
53	0.9250	0.9268	0.9285	0.9303	0.9320	0.9338	0.9355	0.9372	0.9390	0.9407	3	6	9	12	15
54	0.9425	0.9442	0.9460	0.9477	0.9495	0.9512	0.9529	0.9547	0.9564	0.9582	3	6	9	12	15
55	0.9599	0.9617	0.9634	0.9652	0.9669	0.9687	0.9704	0.9721	0.9739	0.9756	3	6	9	12	15
56	0.9774	0.9791	0.9809	0.9826	0.9844	0.9861	0.9879	0.9896	0.9913	0.9931	3	6	9	12	15
57	0.9948	0.9966	0.9983	1.0001	1.0018	1.0036	1.0053	1.0071	1.0088	1.0105	3	6	9	12	15
58	1.0123	1.0140	1.0158	1.0175	1.0193	1.0210	1.0228	1.0245	1.0263	1.0280	3	6	9	12	15
59	1.0297	1.0315	1.0332	1.0350	1.0367	1.0385	1.0402	1.0420	1.0437	1.0455	3	6	9	12	15
60	1.0472	1.0489	1.0507	1.0524	1.0542	1.0559	1.0577	1.0594	1.0612	1.0629	3	6	9	12	15
61	1.0647	1.0664	1.0681	1.0699	1.0716	1.0734	1.0751	1.0769	1.0786	1.0804	3	6	9	12	15
62	1.0821	1.0838	1.0856	1.0873	1.0891	1.0908	1.0926	1.0943	1.0961	1.0978	3	6	9	12	15
63	1.0996	1.1013	1.1030	1.1048	1.1065	1.1083	1.1100	1.1118	1.1135	1.1153	3	6	9	12	15
64	1.1170	1.1188	1.1205	1.1222	1.1240	1.1257	1.1275	1.1292	1.1310	1.1327	3	6	9	12	15
65	1.1345	1.1362	1.1380	1.1397	1.1414	1.1432	1.1449	1.1467	1.1484	1.1502	3	6	9	12	15
66	1.1519	1.1537	1.1554	1.1572	1.1589	1.1606	1.1624	1.1641	1.1659	1.1676	3	6	9	12	15
67	1.1694	1.1711	1.1729	1.1746	1.1764	1.1781	1.1798	1.1816	1.1833	1.1851	3	6	9	12	15
68	1.1868	1.1886	1.1903	1.1921	1.1938	1.1956	1.1973	1.1990	1.2008	1.2025	3	6	9	12	15
69	1.2043	1.2060	1.2078	1.2095	1.2113	1.2130	1.2147	1.2165	1.2182	1.2200	3	6	9	12	15
70	1.2217	1.2235	1.2252	1.2270	1.2287	1.2305	1.2322	1.2339	1.2357	1.2374	3	6	9	12	15
71	1.2392	1.2409	1.2427	1.2444	1.2462	1.2479	1.2497	1.2514	1.2531	1.2549	3	6	9	12	15
72	1.2566	1.2584	1.2601	1.2619	1.2636	1.2654	1.2671	1.2689	1.2706	1.2723	3	6	9	12	15
73	1.2741	1.2758	1.2776	1.2793	1.2811	1.2828	1.2846	1.2863	1.2881	1.2898	3	6	9	12	15
74	1.2915	1.2933	1.2950	1.2968	1.2985	1.3003	1.3020	1.3038	1.3055	1.3073	3	6	9	12	15
75	1.3090	1.3107	1.3125	1.3142	1.3160	1.3177	1.3195	1.3212	1.3230	1.3247	3	6	9	12	15
76	1.3265	1.3282	1.3299	1.3317	1.3334	1.3352	1.3369	1.3387	1.3404	1.3422	3	6	9	12	15
77	1.3439	1.3456	1.3474	1.3491	1.3509	1.3526	1.3544	1.3561	1.3579	1.3596	3	6	9	12	15
78	1.3614	1.3631	1.3648	1.3666	1.3683	1.3701	1.3718	1.3736	1.3753	1.3771	3	6	9	12	15
79	1.3788	1.3806	1.3823	1.3840	1.3858	1.3875	1.3893	1.3910	1.3928	1.3945	3	6	9	12	15
80	1.3963	1.3980	1.3998	1.4015	1.4032	1.4050	1.4067	1.4085	1.4102	1.4120	3	6	9	12	15
81	1.4137	1.4155	1.4172	1.4190	1.4207	1.4224	1.4242	1.4259	1.4277	1.4294	3	6	9	12	15
82	1.4312	1.4329	1.4347	1.4364	1.4382	1.4399	1.4416	1.4434	1.4451	1.4469	3	6	9	12	15
83	1.4486	1.4504	1.4521	1.4539	1.4556	1.4573	1.4591	1.4608	1.4626	1.4643	3	6	9	12	15
84	1.4661	1.4678	1.4696	1.4713	1.4731	1.4748	1.4765	1.4783	1.4800	1.4818	3	6	9	12	15
85	1.4835	1.4853	1.4870	1.4888	1.4905	1.4923	1.4940	1.4957	1.4975	1.4992	3	6	9	12	15
86	1.5010	1.5027	1.5045	1.5062	1.5080	1.5097	1.5115	1.5132	1.5149	1.5167	3	6	9	12	15
87	1.5184	1.5202	1.5219	1.5237	1.5254	1.5272	1.5289	1.5307	1.5324	1.5341	3	6	9	12	15
88	1.5359	1.5376	1.5394	1.5411	1.5429	1.5446	1.5464	1.5481	1.5499	1.5516	3	6	9	12	15
89	1.5533	1.5551	1.5568	1.5586	1.5603	1.5621	1.5638	1.5656	1.5673	1.5691	3	6	9	12	15

35° 46' = 1.4969 radians (direct from table)
0.6219 radians = 35° 38' (by finding 0.6219 in the table and reading off the corresponding angle in degrees)

67

Trigonometrical Ratios for Angles given in Radians

Radians	Degrees	Sine	Cosine	Tangent	Radians	Degrees	Sine	Cosine	Tangent
0.01	0.5730	0.0100	1.0000	0.0100	0.46	26.356	0.4439	0.8961	0.4954
0.02	1.1459	0.0200	0.9998	0.0200	0.47	26.929	0.4529	0.8916	0.5080
0.03	1.7189	0.0300	0.9996	0.0300	0.48	27.502	0.4618	0.8870	0.5206
0.04	2.2918	0.0400	0.9992	0.0400	0.49	28.075	0.4706	0.8823	0.5334
0.05	2.8648	0.0500	0.9988	0.0500	0.50	28.648	0.4794	0.8776	0.5463
0.06	3.4377	0.0600	0.9982	0.0601	0.51	29.221	0.4882	0.8727	0.5594
0.07	4.0107	0.0699	0.9976	0.0701	0.52	29.794	0.4969	0.8678	0.5726
0.08	4.5837	0.0799	0.9968	0.0802	0.53	30.367	0.5055	0.8628	0.5859
0.09	5.1566	0.0899	0.9960	0.0902	0.54	30.940	0.5141	0.8577	0.5994
0.10	5.7296	0.0998	0.9950	0.1003	0.55	31.513	0.5227	0.8525	0.6131
0.11	6.3025	0.1098	0.9940	0.1104	0.56	32.086	0.5312	0.8473	0.6270
0.12	6.8755	0.1197	0.9928	0.1206	0.57	32.659	0.5396	0.8419	0.6410
0.13	7.4485	0.1296	0.9916	0.1307	0.58	33.232	0.5480	0.8365	0.6552
0.14	8.0214	0.1395	0.9902	0.1409	0.59	33.805	0.5564	0.8309	0.6696
0.15	8.5944	0.1494	0.9888	0.1511	0.60	34.377	0.5646	0.8253	0.6841
0.16	9.1673	0.1593	0.9872	0.1614	0.61	34.950	0.5729	0.8196	0.6989
0.17	9.7403	0.1692	0.9856	0.1717	0.62	35.523	0.5810	0.8139	0.7139
0.18	10.313	0.1790	0.9838	0.1820	0.63	36.096	0.5891	0.8080	0.7291
0.19	10.886	0.1889	0.9820	0.1923	0.64	36.669	0.5972	0.8021	0.7445
0.20	11.459	0.1987	0.9801	0.2027	0.65	37.242	0.6052	0.7961	0.7602
0.21	12.032	0.2085	0.9780	0.2131	0.66	37.815	0.6131	0.7900	0.7761
0.22	12.605	0.2182	0.9759	0.2236	0.67	38.388	0.6210	0.7838	0.7923
0.23	13.178	0.2280	0.9737	0.2341	0.68	38.961	0.6288	0.7776	0.8087
0.24	13.751	0.2377	0.9713	0.2447	0.69	39.534	0.6365	0.7712	0.8253
0.25	14.324	0.2474	0.9689	0.2553	0.70	40.107	0.6442	0.7648	0.8423
0.26	14.897	0.2571	0.9664	0.2660	0.71	40.680	0.6518	0.7584	0.8595
0.27	15.470	0.2667	0.9638	0.2768	0.72	41.253	0.6594	0.7518	0.8771
0.28	16.043	0.2764	0.9611	0.2876	0.73	41.826	0.6669	0.7452	0.8949
0.29	16.616	0.2860	0.9582	0:2984	0.74	42.399	0.6743	0.7385	0.9131
0.30	17.189	0.2955	0.9553	0.3093	0.75	42.972	0.6816	0.7317	0.9316
0.31	17.762	0.3051	0.9523	0.3203	0.76	43.545	0.6889	0.7248	0.9505
0.32	18.335	0.3146	0.9492	0.3314	0.77	44.118	0.6961	0.7179	0.9697
0.33	18.908	0.3240	0.9460	0.3425	0.78	44.691	0.7033	0.7109	0.9893
0.34	19.481	0.3335	0.9428	0.3537	0.79	45.264	0.7104	0.7038	1.0092
0.35	20.054	0.3429	0.9394	0.3650	0.80	45.837	0.7174	0.6967	1.0296
0.36	20.626	0.3523	0.9359	0.3764	0.81	46.410	0.7243	0.6895	1.0505
0.37	21.199	0.3616	0.9323	0.3879	0.82	46.983	0.7311	0.6822	1.0717
0.38	21.772	0.3709	0.9287	0.3994	0.83	47.556	0.7379	0.6749	1.0934
0.39	22.345	0.3802	0.9249	0.4111	0.84	48.128	0.7446	0.6675	1.1156
0.40	22.918	0.3894	0.9211	0.4228	0.85	48.701	0.7513	0.6600	1.1383
0.41	23.491	0.3986	0.9171	0.4346	0.86	49.274	0.7578	0.6524	1.1616
0.42	24.064	0.4078	0.9131	0.4466	0.87	49.847	0.7643	0.6448	1.1853
0.43	24.637	0.4169	0.9090	0.4586	0.88	50.420	0.7707	0.6372	1.2097
0.44	25.210	0.4259	0.9048	0.4708	0.89	50.993	0.7771	0.6294	1.2346
0.45	25.783	0.4350	0.9004	0.4831	0.90	51.566	0.7833	0.6216	1.2602

Trigonometrical Ratios for Angles given in Radians

Radians	Degrees	Sine	Cosine	Tangent	Radians	Degrees	Sine	Cosine	Tangent
0.91	52.139	0.7895	0.6137	1.2864	1.26	72.193	0.9521	0.3058	3.1133
0.92	52.712	0.7956	0.6058	1.3133	1.27	72.766	0.9551	0.2963	3.2236
0.93	53.285	0.8016	0.5978	1.3409	1.28	73.339	0.9580	0.2867	3.3414
0.94	53.858	0.8076	0.5898	1.3692	1.29	73.912	0.9608	0.2771	3.4672
0.95	54.431	0.8134	0.5817	1.3984	1.30	74.485	0.9636	0.2675	3.6021
0.96	55.004	0.8192	0.5735	1.4284	1.31	75.058	0.9662	0.2579	3.7471
0.97	55.577	0.8249	0.5653	1.4592	1.32	75.630	0.9687	0.2482	3.9034
0.98	56.150	0.8305	0.5570	1.4910	1.33	76.203	0.9711	0.2385	4.0723
0.99	56.723	0.8360	0.5487	1.5237	1.34	76.776	0.9735	0.2288	4.2556
1.00	57.296	0.8415	0.5403	1.5574	1.35	77.349	0.9757	0.2190	4.4552
1.01	57.869	0.8468	0.5319	1.5922	1.36	77.922	0.9779	0.2092	4.6735
1.02	58.442	0.8521	0.5234	1.6281	1.37	78.495	0.9799	0.1995	4.9131
1.03	59.015	0.8573	0.5148	1.6652	1.38	79.068	0.9819	0.1896	5.1775
1.04	59.588	0.8624	0.5062	1.7036	1.39	79.641	0.9837	0.1798	5.4707
1.05	60.161	0.8674	0.4976	1.7433	1.40	80.214	0.9855	0.1700	5.7979
1.06	60.734	0.8724	0.4889	1.7844	1.41	80.787	0.9871	0.1601	6.1654
1.07	61.307	0.8772	0.4801	1.8270	1.42	81.360	0.9887	0.1502	6.5812
1.08	61.879	0.8820	0.4713	1.8712	1.43	81.933	0.9901	0.1403	7.0555
1.09	62.452	0.8866	0.4625	1.9171	1.44	82.506	0.9915	0.1304	7.6019
1.10	63.025	0.8912	0.4536	1.9648	1.45	83.079	0.9927	0.1205	8.2382
1.11	63.598	0.8957	0.4447	2.0143	1.46	83.652	0.9939	0.1106	8.9887
1.12	64.171	0.9001	0.4357	2.0660	1.47	84.225	0.9949	0.1006	9.8875
1.13	64.744	0.9044	0.4267	2.1198	1.48	84.798	0.9959	0.0907	10.984
1.14	65.317	0.9086	0.4176	2.1759	1.49	85.371	0.9967	0.0807	12.350
1.15	65.890	0.9128	0.4085	2.2345	1.50	85.944	0.9975	0.0707	14.102
1.16	66.463	0.9168	0.3993	2.2958	1.51	86.517	0.9982	0.0608	16.428
1.17	67.036	0.9208	0.3902	2.3600	1.52	87.090	0.9987	0.0508	19.670
1.18	67.609	0.9246	0.3809	2.4273	1.53	87.663	0.9992	0.0408	24.499
1.19	68.182	0.9284	0.3717	2.4979	1.54	88.236	0.9995	0.0308	32.462
1.20	68.755	0.9320	0.3624	2.5722	1.55	88.809	0.9998	0.0208	48.081
1.21	69.328	0.9356	0.3530	2.6503	1.56	89.381	0.9999	0.0108	92.632
1.22	69.901	0.9391	0.3436	2.7328	1.57	89.954	1.0000	0.0008	1257.5
1.23	70.474	0.9425	0.3342	2.8198	$\pi/2$	90.000	1.0000	0.0000	∞
1.24	71.047	0.9458	0.3248	2.9119					
1.25	71.620	0.9490	0.3153	3.0096					

For intermediate values between those given use linear interpolation. Thus for 0.363 radians

$$0.363 \text{ radians} = (21.199 - 20.626) \times \tfrac{3}{10} + 20.626$$
$$= 20.798°$$
$$\sin 0.363 \text{ radians} = (0.3616 - 0.3523) \times \tfrac{3}{10} + 0.3523$$
$$= 0.3551$$
$$\cos 0.363 \text{ radians} = (0.9359 - 0.9323) \times \tfrac{7}{10} + 0.9323$$
$$= 0.9348$$
$$\tan 0.363 \text{ radians} = (0.3879 - 0.3764) \times \tfrac{3}{10} + 0.3764$$
$$= 0.3799$$

Table of Squares

x	0	1	2	3	4	5	6	7	8	9	1	2	3	4	5	6	7	8	9
1.0	1.000	1.020	1.040	1.061	1.082	1.103	1.124	1.145	1.166	1.188	2	4	6	8	10	13	15	17	19
1.1	1.210	1.232	1.254	1.277	1.300	1.323	1.346	1.369	1.392	1.416	2	5	7	9	11	14	16	18	21
1.2	1.440	1.464	1.488	1.513	1.538	1.563	1.588	1.613	1.638	1.664	2	5	7	10	12	15	17	20	22
1.3	1.690	1.716	1.742	1.769	1.796	1.823	1.850	1.877	1.904	1.932	3	5	8	11	13	16	19	22	24
1.4	1.960	1.988	2.016	2.045	2.074	2.103	2.132	2.161	2.190	2.220	3	6	9	12	14	17	20	23	26
1.5	2.250	2.280	2.310	2.341	2.372	2.403	2.434	2.465	2.496	2.528	3	6	9	12	15	19	22	25	28
1.6	2.560	2.592	2.624	2.657	2.690	2.723	2.756	2.789	2.822	2.856	3	7	10	13	16	20	23	26	30
1.7	2.890	2.924	2.958	2.993	3.028	3.063	3.098	3.133	3.168	3.204	3	7	10	14	17	21	24	28	31
1.8	3.240	3.276	3.312	3.349	3.386	3.423	3.460	3.497	3.534	3.572	4	7	11	15	18	22	26	30	33
1.9	3.610	3.648	3.686	3.725	3.764	3.803	3.842	3.881	3.920	3.960	4	8	12	16	19	23	27	31	35
2.0	4.000	4.040	4.080	4.121	4.162	4.203	4.244	4.285	4.326	4.368	4	8	12	16	20	25	29	33	37
2.1	4.410	4.452	4.494	4.537	4.580	4.623	4.666	4.709	4.752	4.796	4	9	13	17	21	26	30	34	39
2.2	4.840	4.884	4.928	4.973	5.018	5.063	5.108	5.153	5.198	5.244	4	9	13	18	22	27	31	36	40
2.3	5.290	5.336	5.382	5.429	5.476	5.523	5.570	5.617	5.664	5.712	5	9	14	19	23	28	33	38	42
2.4	5.760	5.808	5.856	5.905	5.954	6.003	6.052	6.101	6.150	6.200	5	10	15	20	24	29	34	39	44
2.5	6.250	6.300	6.350	6.401	6.452	6.503	6.554	6.605	6.656	6.708	5	10	15	20	25	31	36	41	46
2.6	6.760	6.812	6.864	6.917	6.970	7.023	7.076	7.129	7.182	7.236	5	11	16	21	26	32	37	42	48
2.7	7.290	7.344	7.398	7.453	7.508	7.563	7.618	7.673	7.728	7.784	5	11	16	22	27	33	38	44	49
2.8	7.840	7.896	7.952	8.009	8.066	8.123	8.180	8.237	8.294	8.352	6	11	17	23	28	34	40	46	51
2.9	8.410	8.468	8.526	8.585	8.644	8.703	8.762	8.821	8.880	8.940	6	12	18	24	29	35	41	47	53
3.0	9.000	9.060	9.120	9.181	9.242	9.303	9.364	9.425	9.486	9.548	6	12	18	24	30	37	43	49	55
3.1	9.610	9.672	9.734	9.797	9.860	9.923	9.986	10.05	10.11	10.18	6	13	19	25	31	38	44	50	57
3.2	10.24	10.30	10.37	10.43	10.50	10.56	10.63	10.69	10.76	10.82	1	1	2	3	3	4	5	5	6
3.3	10.89	10.96	11.02	11.09	11.16	11.22	11.29	11.36	11.42	11.49	1	1	2	3	3	4	5	5	6
3.4	11.56	11.63	11.70	11.76	11.83	11.90	11.97	12.04	12.11	12.18	1	1	2	3	3	4	5	6	6
3.5	12.25	12.32	12.39	12.46	12.53	12.60	12.67	12.74	12.82	12.89	1	1	2	3	4	4	5	6	6
3.6	12.96	13.03	13.10	13.18	13.25	13.32	13.40	13.47	13.54	13.62	1	1	2	3	4	4	5	6	7
3.7	13.69	13.76	13.84	13.91	13.99	14.06	14.14	14.21	14.29	14.36	1	2	2	3	4	4	5	6	7
3.8	14.44	14.52	14.59	14.67	14.75	14.82	14.90	14.98	15.05	15.13	1	2	2	3	4	5	5	6	7
3.9	15.21	15.29	15.37	15.44	15.52	15.60	15.68	15.76	15.84	15.92	1	2	2	3	4	5	6	6	7
4.0	16.00	16.08	16.16	16.24	16.32	16.40	16.48	16.56	16.65	16.73	1	2	2	3	4	5	6	6	7
4.1	16.81	16.89	16.97	17.06	17.14	17.22	17.31	17.39	17.47	17.56	1	2	2	3	4	5	6	7	7
4.2	17.64	17.72	17.81	17.89	17.98	18.06	18.15	18.23	18.32	18.40	1	2	3	3	4	5	6	7	8
4.3	18.49	18.58	18.66	18.75	18.84	18.92	19.01	19.10	19.18	19.27	1	2	3	3	4	5	6	7	8
4.4	19.36	19.45	19.54	19.62	19.71	19.80	19.89	19.98	20.07	20.16	1	2	3	4	4	5	6	7	8
4.5	20.25	20.34	20.43	20.52	20.61	20.70	20.79	20.88	20.98	21.07	1	2	3	4	5	5	6	7	8
4.6	21.16	21.25	21.34	21.44	21.53	21.62	21.72	21.81	21.90	22.00	1	2	3	4	5	6	7	7	8
4.7	22.09	22.18	22.28	22.37	22.47	22.56	22.66	22.75	22.85	22.94	1	2	3	4	5	6	7	8	9
4.8	23.04	23.14	23.23	23.33	23.43	23.52	23.62	23.72	23.81	23.91	1	2	3	4	5	6	7	8	9
4.9	24.01	24.11	24.21	24.30	24.40	24.50	24.60	24.70	24.80	24.90	1	2	3	4	5	6	7	8	9
5.0	25.00	25.10	25.20	25.30	25.40	25.50	25.60	25.70	25.81	25.91	1	2	3	4	5	6	7	8	9
5.1	26.01	26.11	26.21	26.32	26.42	26.52	26.63	26.73	26.83	26.94	1	2	3	4	5	6	7	8	9
5.2	27.04	27.14	27.25	27.35	27.46	27.56	27.67	27.77	27.88	27.98	1	2	3	4	5	6	7	8	9
5.3	28.09	28.20	28.30	28.41	28.52	28.62	28.73	28.84	28.94	29.05	1	2	3	4	5	6	7	9	10
5.4	29.16	29.27	29.38	29.48	29.59	29.70	29.81	29.92	30.03	30.14	1	2	3	4	5	7	8	9	10

The table of squares give the squares of numbers from 1 to 10. To find the squares of numbers greater than 10 proceed as follows:

To find $(468.8)^2$

$$(468.8)^2 = (4.688 \times 100)^2 = 4.688^2 \times 100^2 = 21.97 \times 10^4$$
$$\text{or } 219\,700$$

Table of Squares

x	0	1	2	3	4	5	6	7	8	9	1	2	3	4	5	6	7	8	9
5.5	30.25	30.36	30.47	30.58	30.69	30.80	30.91	31.02	31.14	31.25	1	2	3	4	6	7	8	9	10
5.6	31.36	31.47	31.58	31.70	31.81	31.92	32.04	32.15	32.26	32.38	1	2	3	5	6	7	8	9	10
5.7	32.49	32.60	32.72	32.83	32.95	33.06	33.18	33.29	33.41	33.52	1	2	3	5	6	7	8	9	10
5.8	33.64	33.76	33.87	33.99	34.11	34.22	34.34	34.46	34.57	34.69	1	2	4	5	6	7	8	9	11
5.9	34.81	34.93	35.05	35.16	35.28	35.40	35.52	35.64	35.76	35.88	1	2	4	5	6	7	8	10	11
6.0	36.00	36.12	36.24	36.36	36.48	36.60	36.72	36.84	36.97	37.09	1	2	4	5	6	7	9	10	11
6.1	37.21	37.33	37.45	37.58	37.70	37.82	37.95	38.07	38.19	38.32	1	2	4	5	6	7	9	10	11
6.2	38.44	38.56	38.69	38.81	38.94	39.06	39.19	39.31	39.44	39.56	1	3	4	5	6	8	9	10	11
6.3	39.69	39.82	39.94	40.07	40.20	40.32	40.45	40.58	40.70	40.83	1	3	4	5	6	8	9	10	11
6.4	40.96	41.09	41.22	41.34	41.47	41.60	41.73	41.86	41.99	42.12	1	3	4	5	6	8	9	10	12
6.5	42.25	42.38	42.51	42.64	42.77	42.90	43.03	43.16	43.30	43.43	1	3	4	5	7	8	9	10	12
6.6	43.56	43.69	43.82	43.96	44.09	44.22	44.36	44.49	44.62	44.76	1	3	4	5	7	8	9	11	12
6.7	44.89	45.02	45.16	45.29	45.43	45.56	45.70	45.83	45.97	46.10	1	3	4	5	7	8	9	11	12
6.8	46.24	46.38	46.51	46.65	46.79	46.92	47.06	47.20	47.33	47.47	1	3	4	5	7	8	10	11	12
6.9	47.61	47.75	47.89	48.02	48.16	48.30	48.44	48.58	48.72	48.86	1	3	4	6	7	8	10	11	13
7.0	49.00	49.14	49.28	49.42	49.56	49.70	49.84	49.98	50.13	50.27	1	3	4	6	7	8	10	11	13
7.1	50.41	50.55	50.69	50.84	50.98	51.12	51.27	51.41	51.55	51.70	1	3	4	6	7	9	10	11	13
7.2	51.84	51.98	52.13	52.27	52.42	52.56	52.71	52.85	53.00	53.14	1	3	4	6	7	9	10	12	13
7.3	53.29	53.44	53.58	53.73	53.88	54.02	54.17	54.32	54.46	54.61	1	3	4	6	7	9	10	12	13
7.4	54.76	54.91	55.06	55.20	55.35	55.50	55.65	55.80	55.95	56.10	1	3	4	6	7	9	10	12	13
7.5	56.25	56.40	56.55	56.70	56.85	57.00	57.15	57.30	57.46	57.61	2	3	5	6	8	9	11	12	14
7.6	57.76	57.91	58.06	58.22	58.37	58.52	58.68	58.83	58.98	59.14	2	3	5	6	8	9	11	12	14
7.7	59.29	59.44	59.60	59.75	59.91	60.06	60.22	60.37	60.53	60.68	2	3	5	6	8	9	11	13	14
7.8	60.84	61.00	61.15	61.31	61.47	61.62	61.78	61.94	62.09	62.25	2	3	5	6	8	9	11	13	14
7.9	62.41	62.57	62.73	62.88	63.04	63.20	63.36	63.52	63.68	63.84	2	3	5	6	8	10	11	13	14
8.0	64.00	64.16	64.32	64.48	64.64	64.80	64.96	65.12	65.29	65.45	2	3	5	6	8	10	11	13	14
8.1	65.61	65.77	65.93	66.10	66.26	66.42	67.59	66.75	66.91	67.08	2	3	5	7	8	10	11	13	15
8.2	67.24	67.40	67.57	67.73	67.90	68.06	68.23	68.39	68.56	68.72	2	3	5	7	8	10	12	13	15
8.3	68.89	69.06	69.22	69.39	69.56	69.72	69.89	70.06	70.22	70.39	2	3	5	7	8	10	12	13	15
8.4	70.56	70.73	70.90	71.06	71.23	71.40	71.57	71.74	71.91	72.08	2	3	5	7	8	10	12	14	15
8.5	72.25	72.42	72.59	72.76	72.93	73.10	73.27	73.44	73.62	73.79	2	3	5	7	9	10	12	14	15
8.6	73.96	74.13	74.30	74.48	74.65	74.82	75.00	75.17	75.34	75.52	2	3	5	7	9	10	12	14	16
8.7	75.69	75.86	76.04	76.21	76.39	76.56	76.74	76.91	77.09	77.26	2	4	5	7	9	11	12	14	16
8.8	77.44	77.62	77.79	77.97	78.15	78.32	78.50	78.68	78.85	79.03	2	4	5	7	9	11	12	14	16
8.9	79.21	79.39	79.57	79.74	79.92	80.10	80.28	80.46	80.64	80.82	2	4	5	7	9	11	13	14	16
9.0	81.00	81.18	81.36	81.54	81.72	81.90	82.08	82.26	82.45	82.63	2	4	5	7	9	11	13	14	16
9.1	82.81	82.99	83.17	83.36	83.54	83.72	83.91	84.09	84.27	84.46	2	4	5	7	9	11	13	15	16
9.2	84.64	84.82	85.01	85.19	85.38	85.56	85.75	85.93	86.12	86.30	2	4	6	7	9	11	13	15	17
9.3	86.49	86.68	86.86	87.05	87.24	87.42	87.61	87.80	87.98	88.17	2	4	6	7	9	11	13	15	17
9.4	88.36	88.55	88.74	88.92	89.11	89.30	89.49	89.68	89.87	90.06	2	4	6	8	9	11	13	15	17
9.5	90.25	90.44	90.63	90.82	91.01	91.20	91.39	91.58	91.78	91.97	2	4	6	8	10	11	13	15	17
9.6	92.16	92.35	92.54	92.74	92.93	93.12	93.32	93.51	93.70	93.90	2	4	6	8	10	12	14	15	17
9.7	94.09	94.28	94.48	94.67	94.87	95.06	95.26	95.45	95.65	95.84	2	4	6	8	10	12	14	16	18
9.8	96.04	96.24	96.43	96.63	96.83	97.02	97.22	97.42	97.61	97.81	2	4	6	8	10	12	14	16	18
9.9	98.01	98.21	98.41	98.60	98.80	99.00	99.20	99.40	99.60	99.80	2	4	6	8	10	12	14	16	18

To find $(0.2388)^2$

$$(0.2388)^2 = (2.388 \times \tfrac{1}{10})^2 = (2.388)^2 \times (\tfrac{1}{10})^2$$
$$= 5.703 \times \tfrac{1}{100} = 5.703 \times 10^{-2} \text{ or } 0.057\,03$$

Table of Square Roots from 1–10

	0	1	2	3	4	5	6	7	8	9	1	2	3	4	5	6	7	8	9
1.0	1.000	1.005	1.010	1.015	1.020	1.025	1.030	1.034	1.039	1.044	0	1	1	2	2	3	3	4	4
1.1	1.049	1.054	1.058	1.063	1.068	1.072	1.077	1.082	1.086	1.091	0	1	1	2	2	3	3	4	4
1.2	1.095	1.100	1.105	1.109	1.114	1.118	1.122	1.127	1.131	1.136	0	1	1	2	2	3	3	4	4
1.3	1.140	1.145	1.149	1.153	1.158	1.162	1.166	1.170	1.175	1.179	0	1	1	2	2	3	3	3	4
1.4	1.183	1.187	1.192	1.196	1.200	1.204	1.208	1.212	1.217	1.221	0	1	1	2	2	3	3	3	4
1.5	1.225	1.229	1.233	1.237	1.241	1.245	1.249	1.253	1.257	1.261	0	1	1	2	2	2	3	3	4
1.6	1.265	1.269	1.273	1.277	1.281	1.285	1.288	1.292	1.296	1.300	0	1	1	2	2	2	3	3	3
1.7	1.304	1.308	1.311	1.315	1.319	1.323	1.327	1.330	1.334	1.338	0	1	1	1	2	2	3	3	3
1.8	1.342	1.345	1.349	1.353	1.356	1.360	1.364	1.367	1.371	1.375	0	1	1	1	2	2	3	3	3
1.9	1.378	1.382	1.386	1.389	1.393	1.396	1.400	1.404	1.407	1.411	0	1	1	1	2	2	3	3	3
2.0	1.414	1.418	1.421	1.425	1.428	1.432	1.435	1.439	1.442	1.446	0	1	1	1	2	2	2	3	3
2.1	1.449	1.453	1.456	1.459	1.463	1.466	1.470	1.473	1.476	1.480	0	1	1	1	2	2	2	3	3
2.2	1.483	1.487	1.490	1.493	1.497	1.500	1.503	1.507	1.510	1.513	0	1	1	1	2	2	2	3	3
2.3	1.517	1.520	1.523	1.526	1.530	1.533	1.536	1.539	1.543	1.546	0	1	1	1	2	2	2	2	3
2.4	1.549	1.552	1.556	1.559	1.562	1.565	1.568	1.572	1.575	1.578	0	1	1	1	2	2	2	2	3
2.5	1.581	1.584	1.587	1.591	1.594	1.597	1.600	1.603	1.606	1.609	0	1	1	1	2	2	2	2	3
2.6	1.612	1.616	1.619	1.622	1.625	1.628	1.631	1.634	1.637	1.640	0	1	1	1	2	2	2	2	3
2.7	1.643	1.646	1.649	1.652	1.655	1.658	1.661	1.664	1.667	1.670	0	1	1	1	2	2	2	2	3
2.8	1.673	1.676	1.679	1.682	1.685	1.688	1.691	1.694	1.697	1.700	0	1	1	1	2	2	2	2	3
2.9	1.703	1.706	1.709	1.712	1.715	1.718	1.720	1.723	1.726	1.729	0	1	1	1	1	2	2	2	3
3.0	1.732	1.735	1.738	1.741	1.744	1.746	1.749	1.752	1.755	1.758	0	1	1	1	1	2	2	2	3
3.1	1.761	1.764	1.766	1.769	1.772	1.775	1.778	1.780	1.783	1.786	0	1	1	1	1	2	2	2	3
3.2	1.789	1.792	1.794	1.797	1.800	1.803	1.806	1.808	1.811	1.814	0	1	1	1	1	2	2	2	2
3.3	1.817	1.819	1.822	1.825	1.828	1.830	1.833	1.836	1.838	1.841	0	1	1	1	1	2	2	2	2
3.4	1.844	1.847	1.849	1.852	1.855	1.857	1.860	1.863	1.865	1.868	0	1	1	1	1	2	2	2	2
3.5	1.871	1.873	1.876	1.879	1.881	1.884	1.887	1.889	1.892	1.895	0	1	1	1	1	2	2	2	2
3.6	1.897	1.900	1.903	1.905	1.908	1.910	1.913	1.916	1.918	1.921	0	1	1	1	1	2	2	2	2
3.7	1.924	1.926	1.929	1.931	1.934	1.936	1.939	1.942	1.944	1.947	0	1	1	1	1	2	2	2	2
3.8	1.949	1.952	1.954	1.957	1.960	1.962	1.965	1.967	1.970	1.972	0	1	1	1	1	2	2	2	2
3.9	1.975	1.977	1.980	1.982	1.985	1.987	1.990	1.992	1.995	1.997	0	0	1	1	1	1	2	2	2
4.0	2.000	2.002	2.005	2.007	2.010	2.012	2.015	2.017	2.020	2.022	0	0	1	1	1	1	2	2	2
4.1	2.025	2.027	2.030	2.032	2.035	2.037	2.040	2.042	2.045	2.047	0	0	1	1	1	1	2	2	2
4.2	2.049	2.052	2.054	2.057	2.059	2.062	2.064	2.066	2.069	2.071	0	0	1	1	1	1	2	2	2
4.3	2.074	2.076	2.078	2.081	2.083	2.086	2.088	2.090	2.093	2.095	0	0	1	1	1	1	1	2	2
4.4	2.098	2.100	2.102	2.105	2.107	2.110	2.112	2.114	2.117	2.119	0	0	1	1	1	1	1	2	2
4.5	2.121	2.124	2.126	2.128	2.131	2.133	2.135	2.138	2.140	2.142	0	0	1	1	1	1	2	2	2
4.6	2.145	2.147	2.149	2.152	2.154	2.156	2.159	2.161	2.163	2.166	0	0	1	1	1	1	2	2	2
4.7	2.168	2.170	2.173	2.175	2.177	2.179	2.182	2.184	2.186	2.189	0	0	1	1	1	1	2	2	2
4.8	2.191	2.193	2.195	2.198	2.200	2.202	2.205	2.207	2.209	2.211	0	0	1	1	1	1	2	2	2
4.9	2.214	2.216	2.218	2.220	2.223	2.225	2.227	2.229	2.232	2.234	0	0	1	1	1	1	2	2	2
5.0	2.236	2.238	2.241	2.243	2.245	2.247	2.249	2.252	2.254	2.256	0	0	1	1	1	1	2	2	2
5.1	2.258	2.261	2.263	2.265	2.267	2.269	2.272	2.274	2.276	2.278	0	0	1	1	1	1	2	2	2
5.2	2.280	2.283	2.285	2.287	2.289	2.291	2.293	2.296	2.298	2.300	0	0	1	1	1	1	2	2	2
5.3	2.302	2.304	2.307	2.309	2.311	2.313	2.315	2.317	2.319	2.322	0	0	1	1	1	1	2	2	2
5.4	2.324	2.326	2.328	2.330	2.332	2.335	2.337	2.339	2.341	2.343	0	0	1	1	1	1	2	2	2

There are two sets of square root tables one giving square roots of numbers between 1 and 10 and the other, square roots of numbers between 10 and 100.

$$\sqrt{1.873} = 1.368 \text{ (using square roots of numbers between 1 and 10)}$$
$$\sqrt{18.73} = 4.327 \text{ (using square roots of numbers between 10 and 100)}$$

Table of Square Roots from 1–10

	0	1	2	3	4	5	6	7	8	9	1	2	3	4	5	6	7	8	9
5.5	2.345	2.347	2.349	2.352	2.354	2.356	2.358	2.360	2.362	2.364	0	0	1	1	1	1	1	2	2
5.6	2.366	2.369	2.371	2.373	2.375	2.377	2.379	2.381	2.383	2.385	0	0	1	1	1	1	1	2	2
5.7	2.387	2.390	2.392	2.394	2.396	2.398	2.400	2.402	2.404	2.406	0	0	1	1	1	1	1	2	2
5.8	2.408	2.410	2.412	2.415	2.417	2.419	2.421	2.423	2.425	2.427	0	0	1	1	1	1	1	2	2
5.9	2.429	2.431	2.433	2.435	2.437	2.439	2.441	2.443	2.445	2.447	0	0	1	1	1	1	1	2	2
6.0	2.449	2.452	2.454	2.456	2.458	2.460	2.462	2.464	2.466	2.468	0	0	1	1	1	1	1	2	2
6.1	2.470	2.472	2.474	2.476	2.478	2.480	2.482	2.484	2.486	2.488	0	0	1	1	1	1	1	2	2
6.2	2.490	2.492	2.494	2.496	2.498	2.500	2.502	2.504	2.506	2.508	0	0	1	1	1	1	1	2	2
6.3	2.510	2.512	2.514	2.516	2.518	2.520	2.522	2.524	2.526	2.528	0	0	1	1	1	1	1	2	2
6.4	2.530	2.532	2.534	2.536	2.538	2.540	2.542	2.544	2.546	2.548	0	0	1	1	1	1	1	2	2
6.5	2.550	2.551	2.553	2.555	2.557	2.559	2.561	2.563	2.565	2.567	0	0	1	1	1	1	1	2	2
6.6	2.569	2.571	2.573	2.575	2.577	2.579	2.581	2.583	2.585	2.587	0	0	1	1	1	1	1	2	2
6.7	2.588	2.590	2.592	2.594	2.596	2.598	2.600	2.602	2.604	2.606	0	0	1	1	1	1	1	2	2
6.8	2.608	2.610	2.612	2.613	2.615	2.617	2.619	2.621	2.623	2.625	0	0	1	1	1	1	1	2	2
6.9	2.627	2.629	2.631	2.632	2.634	2.636	2.638	2.640	2.642	2.644	0	0	1	1	1	1	1	2	2
7.0	2.646	2.648	2.650	2.651	2.653	2.655	2.657	2.659	2.661	2.663	0	0	1	1	1	1	1	2	2
7.1	2.665	2.666	2.668	2.670	2.672	2.674	2.676	2.678	2.680	2.681	0	0	1	1	1	1	1	2	2
7.2	2.683	2.685	2.687	2.689	2.691	2.693	2.694	2.696	2.698	2.700	0	0	1	1	1	1	1	2	2
7.3	2.702	2.704	2.706	2.707	2.709	2.711	2.713	2.715	2.717	2.718	0	0	1	1	1	1	1	2	2
7.4	2.720	2.722	2.724	2.726	2.728	2.729	2.731	2.733	2.735	2.737	0	0	1	1	1	1	1	2	2
7.5	2.739	2.740	2.742	2.744	2.746	2.748	2.750	2.751	2.753	2.755	0	0	1	1	1	1	1	2	2
7.6	2.757	2.759	2.760	2.762	2.764	2.766	2.768	2.769	2.771	2.773	0	0	1	1	1	1	1	2	2
7.7	2.775	2.777	2.778	2.780	2.782	2.784	2.786	2.787	2.789	2.791	0	0	1	1	1	1	1	2	2
7.8	2.793	2.795	2.796	2.798	2.800	2.802	2.804	2.805	2.807	2.809	0	0	1	1	1	1	1	2	2
7.9	2.811	2.812	2.814	2.816	2.818	2.820	2.821	2.823	2.825	2.827	0	0	1	1	1	1	1	2	2
8.0	2.828	2.830	2.832	2.834	2.835	2.837	2.839	2.841	2.843	2.844	0	0	1	1	1	1	1	2	2
8.1	2.846	2.848	2.850	2.851	2.853	2.855	2.857	2.858	2.860	2.862	0	0	1	1	1	1	1	2	2
8.2	2.864	2.865	2.867	2.869	2.871	2.872	2.874	2.876	2.877	2.879	0	0	1	1	1	1	1	2	2
8.3	2.881	2.883	2.884	2.886	2.888	2.890	2.891	2.893	2.895	2.897	0	0	1	1	1	1	1	2	2
8.4	2.898	2.900	2.902	2.903	2.905	2.907	2.909	2.910	2.912	2.914	0	0	1	1	1	1	1	2	2
8.5	2.915	2.917	2.919	2.921	2.922	2.924	2.926	2.927	2.929	2.931	0	0	1	1	1	1	1	2	2
8.6	2.933	2.934	2.936	2.938	2.939	2.941	2.943	2.944	2.946	2.948	0	0	1	1	1	1	1	2	2
8.7	2.950	2.951	2.953	2.955	2.956	2.958	2.960	2.961	2.963	2.965	0	0	1	1	1	1	1	2	2
8.8	2.966	2.968	2.970	2.972	2.973	2.975	2.977	2.978	2.980	2.982	0	0	1	1	1	1	1	2	2
8.9	2.983	2.985	2.987	2.988	2.990	2.992	2.993	2.995	2.997	2.998	0	0	1	1	1	1	1	2	2
9.0	3.000	3.002	3.003	3.005	3.007	3.008	3.010	3.012	3.013	3.015	0	0	0	1	1	1	1	1	1
9.1	3.017	3.018	3.020	3.022	3.023	3.025	3.027	3.028	3.030	3.032	0	0	0	1	1	1	1	1	1
9.2	3.033	3.035	3.036	3.038	3.040	3.041	3.043	3.045	3.046	3.048	0	0	0	1	1	1	1	1	1
9.3	3.050	3.051	3.053	3.055	3.056	3.058	3.059	3.061	3.063	3.064	0	0	0	1	1	1	1	1	1
9.4	3.066	3.068	3.069	3.071	3.072	3.074	3.076	3.077	3.079	3.081	0	0	0	1	1	1	1	1	1
9.5	3.082	3.084	3.085	3.087	3.089	3.090	3.092	3.094	3.095	3.097	0	0	0	1	1	1	1	1	1
9.6	3.098	3.100	3.102	3.103	3.105	3.106	3.108	3.110	3.111	3.113	0	0	0	1	1	1	1	1	1
9.7	3.114	3.116	3.118	3.119	3.121	3.122	3.124	3.126	3.127	3.129	0	0	0	1	1	1	1	1	1
9.8	3.130	3.132	3.134	3.135	3.137	3.138	3.140	3.142	3.143	3.145	0	0	0	1	1	1	1	1	1
9.9	3.146	3.148	3.150	3.151	3.153	3.154	3.156	3.158	3.159	3.161	0	0	0	1	1	1	1	1	1

To find $\sqrt{836.3}$

Mark off figures in pairs to the left of the decimal point. Thus 836.3 becomes 8'36.3. The first period is 8 so look up $\sqrt{8.363} = 2.892$. For each period to the left of the decimal there will be one figure to the left of the decimal point in the answer. Thus $\sqrt{836.3} = 28.92$.

To find $\sqrt{8363}$.

Marking off in pairs 8363 becomes 83'63 so look up $\sqrt{83.63} = 9.145$ and hence $\sqrt{8363} = 91.45$.

Table of Square Roots from 10–100

x	0	1	2	3	4	5	6	7	8	9	1	2	3	4	5	6	7	8	9
10	3.162	3.178	3.194	3.209	3.225	3.240	3.256	3.271	3.286	3.302	2	3	5	6	8	10	11	13	14
11	3.317	3.332	3.347	3.362	3.376	3.391	3.406	3.421	3.435	3.450	1	3	4	6	7	9	10	12	13
12	3.464	3.479	3.493	3.507	3.521	3.536	3.550	3.564	3.578	3.592	1	3	4	6	7	8	10	11	13
13	3.606	3.619	3.633	3.647	3.661	3.674	3.688	3.701	3.715	3.728	1	3	4	5	7	8	10	11	12
14	3.742	3.755	3.768	3.782	3.795	3.808	3.821	3.834	3.847	3.860	1	3	4	5	7	8	9	10	12
15	3.873	3.886	3.899	3.912	3.924	3.937	3.950	3.962	3.975	3.987	1	3	4	5	6	8	9	10	11
16	4.000	4.012	4.025	4.037	4.050	4.062	4.074	4.087	4.099	4.111	1	2	4	5	6	7	9	10	11
17	4.123	4.135	4.147	4.159	4.171	4.183	4.195	4.207	4.219	4.231	1	2	4	5	6	7	8	10	11
18	4.243	4.254	4.266	4.278	4.290	4.301	4.313	4.324	4.336	4.347	1	2	3	5	6	7	8	9	10
19	4.359	4.370	4.382	4.393	4.405	4.416	4.427	4.438	4.450	4.461	1	2	3	5	6	7	8	9	10
20	4.472	4.483	4.494	4.506	4.517	4.528	4.539	4.550	4.561	4.572	1	2	3	4	6	7	8	9	10
21	4.583	4.593	4.604	4.615	4.626	4.637	4.648	4.658	4.669	4.680	1	2	3	4	5	6	7	9	10
22	4.690	4.701	4.712	4.722	4.733	4.743	4.754	4.764	4.775	4.785	1	2	3	4	5	6	7	8	10
23	4.796	4.806	4.817	4.827	4.837	4.848	4.858	4.868	4.879	4.889	1	2	3	4	5	6	7	8	9
24	4.899	4.909	4.919	4.930	4.940	4.950	4.960	4.970	4.980	4.990	1	2	3	4	5	6	7	8	9
25	5.000	5.010	5.020	5.030	5.040	5.050	5.060	5.070	5.079	5.089	1	2	3	4	5	6	7	8	9
26	5.099	5.109	5.119	5.128	5.138	5.148	5.158	5.167	5.177	5.187	1	2	3	4	5	6	7	8	9
27	5.196	5.206	5.215	5.225	5.235	5.244	5.254	5.263	5.273	5.282	1	2	3	4	5	6	7	8	9
28	5.292	5.301	5.310	5.320	5.329	5.339	5.348	5.357	5.367	5.376	1	2	3	4	5	6	7	7	8
29	5.385	5.394	5.404	5.413	5.422	5.431	5.441	5.450	5.459	5.468	1	2	3	4	5	5	6	7	8
30	5.477	5.486	5.495	5.505	5.514	5.523	5.532	5.541	5.550	5.559	1	2	3	4	5	5	6	7	8
31	5.568	5.577	5.586	5.595	5.604	5.612	5.621	5.630	5.639	5.648	1	2	3	4	4	5	6	7	8
32	5.657	5.666	5.675	5.683	5.692	5.701	5.710	5.718	5.727	5.736	1	2	3	3	4	5	6	7	8
33	5.745	5.753	5.762	5.771	5.779	5.788	5.797	5.805	5.814	5.822	1	2	3	3	4	5	6	7	8
34	5.831	5.840	5.848	5.857	5.865	5.874	5.882	5.891	5.899	5.908	1	2	3	3	4	5	6	7	8
35	5.916	5.925	5.933	5.941	5.950	5.958	5.967	5.975	5.983	5.992	1	2	2	3	4	5	6	7	8
36	6.000	6.008	6.017	6.025	6.033	6.042	6.050	6.058	6.066	6.075	1	2	2	3	4	5	6	7	7
37	6.083	6.091	6.099	6.107	6.116	6.124	6.132	6.140	6.148	6.156	1	2	2	3	4	5	6	6	7
38	6.164	6.173	6.181	6.189	6.197	6.205	6.213	6.221	6.229	6.237	1	2	2	3	4	5	6	6	7
39	6.245	6.253	6.261	6.269	6.277	6.285	6.293	6.301	6.309	6.317	1	2	2	3	4	5	6	6	7
40	6.325	6.332	6.340	6.348	6.356	6.364	6.372	6.380	6.387	6.395	1	2	2	3	4	5	6	6	7
41	6.403	6.411	6.419	6.427	6.434	6.442	6.450	6.458	6.465	6.473	1	2	2	3	4	5	5	6	7
42	6.481	6.488	6.496	6.504	6.512	6.519	6.527	6.535	6.542	6.550	1	2	2	3	4	5	5	6	7
43	6.557	6.565	6.573	6.580	6.588	6.595	6.603	6.611	6.618	6.626	1	2	2	3	4	5	5	6	7
44	6.633	6.641	6.648	6.656	6.663	6.671	6.678	6.686	6.693	6.701	1	2	2	3	4	4	5	6	7
45	6.708	6.716	6.723	6.731	6.738	6.745	6.753	6.760	6.768	6.775	1	1	2	3	4	4	5	6	7
46	6.782	6.790	6.797	6.804	6.812	6.819	6.826	6.834	6.841	6.848	1	1	2	3	4	4	5	6	7
47	6.856	6.863	6.870	6.877	6.885	6.892	6.899	6.907	6.914	6.921	1	1	2	3	4	4	5	6	6
48	6.928	6.935	6.943	6.950	6.957	6.964	6.971	6.979	6.986	6.993	1	1	2	3	4	4	5	6	6
49	7.000	7.007	7.014	7.021	7.029	7.036	7.043	7.050	7.057	7.064	1	1	2	3	4	4	5	6	6
50	7.071	7.078	7.085	7.092	7.099	7.106	7.113	7.120	7.127	7.134	1	1	2	3	4	4	5	6	6
51	7.141	7.148	7.155	7.162	7.169	7.176	7.183	7.190	7.197	7.204	1	1	2	3	4	4	5	6	6
52	7.211	7.218	7.225	7.232	7.239	7.246	7.253	7.259	7.266	7.273	1	1	2	3	3	4	5	6	6
53	7.280	7.287	7.294	7.301	7.308	7.314	7.321	7.328	7.335	7.342	1	1	2	3	3	4	5	5	6
54	7.348	7.355	7.362	7.369	7.376	7.382	7.389	7.396	7.403	7.409	1	1	2	3	3	4	5	5	6

To find $\sqrt{0.000\,2168}$

In the case of decimal numbers mark off in pairs to the right of the decimal point. Thus 0.02 168 becomes 0.00'02'16'8. Apart from the zero pair the first pair is 02 so look up $\sqrt{2.168} = 1.473$. For each zero pair in the original number there will be one zero following the decimal point in the answer.

Therefore $\sqrt{0.000\,2168} = 0.014\,73$.

74

Table of Square Roots from 10–100

x	0	1	2	3	4	5	6	7	8	9	1	2	3	4	5	6	7	8	9
55	7.416	7.423	7.430	7.436	7.443	7.450	7.457	7.463	7.470	7.477	1	1	2	3	3	4	5	5	6
56	7.483	7.490	7.497	7.503	7.510	7.517	7.523	7.530	7.537	7.543	1	1	2	3	3	4	5	5	6
57	7.550	7.556	7.563	7.570	7.576	7.583	7.589	7.596	7.603	7.609	1	1	2	3	3	4	5	5	6
58	7.616	7.622	7.629	7.635	7.642	7.649	7.655	7.662	7.668	7.675	1	1	2	3	3	4	5	5	6
59	7.681	7.688	7.694	7.701	7.707	7.714	7.720	7.727	7.733	7.740	1	1	2	3	3	4	5	5	6
60	7.746	7.752	7.759	7.765	7.772	7.778	7.785	7.791	7.797	7.804	1	1	2	3	3	4	4	5	6
61	7.810	7.817	7.823	7.829	7.836	7.842	7.849	7.855	7.861	7.868	1	1	2	2	3	4	4	5	5
62	7.874	7.880	7.887	7.893	7.899	7.906	7.912	7.918	7.925	7.931	1	1	2	2	3	4	4	5	5
63	7.937	7.944	7.950	7.956	7.962	7.969	7.975	7.981	7.987	7.994	1	1	2	2	3	4	4	5	5
64	8.000	8.006	8.012	8.019	8.025	8.031	8.037	8.044	8.050	8.056	1	1	2	2	3	4	4	5	5
65	8.062	8.068	8.075	8.081	8.087	8.093	8.099	8.106	8.112	8.118	1	1	2	2	3	4	4	5	5
66	8.124	8.130	8.136	8.142	8.149	8.155	8.161	8.167	8.173	8.179	1	1	2	2	3	4	4	5	5
67	8.185	8.191	8.198	8.204	8.210	8.216	8.222	8.228	8.234	8.240	1	1	2	2	3	4	4	5	5
68	8.246	8.252	8.258	8.264	8.270	8.276	8.283	8.289	8.295	8.301	1	1	2	2	3	4	4	5	5
69	8.307	8.313	8.319	8.325	8.331	8.337	8.343	8.349	8.355	8.361	1	1	2	2	3	4	4	5	5
70	8.367	8.373	8.379	8.385	8.390	8.396	8.402	8.408	8.414	8.420	1	1	2	2	3	4	4	5	5
71	8.426	8.432	8.438	8.444	8.450	8.456	8.462	8.468	8.473	8.479	1	1	2	2	3	3	4	5	5
72	8.485	8.491	8.497	8.503	8.509	8.515	8.521	8.526	8.532	8.538	1	1	2	2	3	3	4	5	5
73	8.544	8.550	8.556	8.562	8.567	8.573	8.579	8.585	8.591	8.597	1	1	2	2	3	4	4	5	5
74	8.602	8.608	8.614	8.620	8.626	8.631	8.637	8.643	8.649	8.654	1	1	2	2	3	3	4	5	5
75	8.660	8.666	8.672	8.678	8.683	8.689	8.695	8.701	8.706	8.712	1	1	2	2	3	3	4	4	5
76	8.718	8.724	8.729	8.735	8.741	8.746	8.752	8.758	8.764	8.769	1	1	2	2	3	3	4	4	5
77	8.775	8.781	8.786	8.792	8.798	8.803	8.809	8.815	8.820	8.826	1	1	2	2	3	3	4	4	5
78	8.832	8.837	8.843	8.849	8.854	8.860	8.866	8.871	8.877	8.883	1	1	2	2	3	3	4	4	5
79	8.888	8.894	8.899	8.905	8.911	8.916	8.922	8.927	8.933	8.939	1	1	2	2	3	3	4	4	5
80	8.944	8.950	8.955	8.961	8.967	8.972	8.978	8.983	8.989	8.994	1	1	2	2	3	3	4	4	5
81	9.000	9.006	9.011	9.017	9.022	9.028	9.033	9.039	9.044	9.050	1	1	2	2	3	3	4	4	5
82	9.055	9.061	9.066	9.072	9.077	9.083	9.088	9.094	9.099	9.105	1	1	2	2	3	3	4	4	5
83	9.110	9.116	9.121	9.127	9.132	9.138	9.143	9.149	9.154	9.160	1	1	2	2	3	3	4	4	5
84	9.165	9.171	9.176	9.182	9.187	9.192	9.198	9.203	9.209	9.214	1	1	2	2	3	3	4	4	5
85	9.220	9.225	9.230	9.236	9.241	9.247	9.252	9.257	9.263	9.268	1	1	2	2	3	3	4	4	5
86	9.274	9.279	9.284	9.290	9.295	9.301	9.306	9.311	9.317	9.322	1	1	2	2	3	3	4	4	5
87	9.327	9.333	9.338	9.343	9.349	9.354	9.359	9.365	9.370	9.375	1	1	2	2	3	3	4	4	5
88	9.381	9.386	9.391	9.397	9.402	9.407	9.413	9.418	9.423	9.429	1	1	2	2	3	3	4	4	5
89	9.434	9.439	9.445	9.450	9.455	9.460	9.466	9.471	9.476	9.482	1	1	2	2	3	3	4	4	5
90	9.487	9.492	9.497	9.503	9.508	9.513	9.518	9.524	9.529	9.534	1	1	2	2	3	3	4	4	5
91	9.539	9.545	9.550	9.555	9.560	9.566	9.571	9.576	9.581	9.586	1	1	2	2	3	3	4	4	5
92	9.592	9.597	9.602	9.607	9.612	9.618	9.623	9.628	9.633	9.638	1	1	2	2	3	3	4	4	5
93	9.644	9.649	9.654	9.659	9.664	9.670	9.675	9.680	9.685	9.690	1	1	2	2	3	3	4	4	5
94	9.695	9.701	9.706	9.711	9.716	9.721	9.726	9.731	9.737	9.742	1	1	2	2	3	3	4	4	5
95	9.747	9.752	9.757	9.762	9.767	9.772	9.778	9.783	9.788	9.793	1	1	2	2	3	3	4	4	5
96	9.798	9.803	9.808	9.813	9.818	9.823	9.829	9.834	9.839	9.844	1	1	2	2	3	3	4	4	5
97	9.849	9.854	9.859	9.864	9.869	9.874	9.879	9.884	9.889	9.894	1	1	2	2	3	3	4	4	5
98	9.899	9.905	9.910	9.915	9.920	9.925	9.930	9.935	9.940	9.945	1	1	2	2	3	3	4	4	5
99	9.950	9.955	9.960	9.965	9.970	9.975	9.980	9.985	9.990	9.995	0	1	1	2	2	3	4	4	4

To find $\sqrt{0.002\,168}$

Marking off in pairs $0.002\,168$ becomes $0.00'21'68$ so look up $\sqrt{21.68} = 4.657$
and hence $\sqrt{0.002\,168} = 0.046\,57$.

Cubes

x	0	1	2	3	4	5	6	7	8	9	Add 1	2	3	4	5	6	7	8	9
1.0	1.000	1.030	1.061	1.093	1.125	1.158	1.191	1.225	1.260	1.295	3	7	10	13	16	20	23	26	30
1.1	1.331	1.368	1.405	1.443	1.482	1.521	1.561	1.602	1.643	1.685	4	8	12	16	20	24	28	32	36
1.2	1.728	1.772	1.816	1.861	1.907	1.953	2.000	2.048	2.097	2.147	5	9	14	19	23	28	33	37	42
1.3	2.197	2.248	2.300	2.353	2.406	2.460	2.515	2.571	2.628	2.686	5	11	16	22	27	33	38	44	49
1.4	2.744	2.803	2.863	2.924	2.986	3.049	3.112	3.177	3.242	3.308	6	13	19	25	31	38	44	50	57
1.5	3.375	3.443	3.512	3.582	3.652	3.724	3.796	3.870	3.944	4.020	7	14	22	29	36	43	50	58	65
1.6	4.096	4.173	4.252	4.331	4.411	4.492	4.574	4.657	4.742	4.827	8	16	24	33	41	49	57	65	73
1.7	4.913	5.000	5.088	5.178	5.268	5.359	5.452	5.545	5.640	5.735	9	18	27	37	46	55	64	73	83
1.8	5.832	5.930	6.029	6.128	6.230	6.332	6.435	6.539	6.645	6.751	10	20	31	41	51	61	72	82	92
1.9	6.859	6.968	7.078	7.189	7.301	7.415	7.530	7.645	7.762	7.881	11	23	34	45	57	68	80	91	103
2.0	8.000	8.121	8.242	8.365	8.490	8.615	8.742	8.870	8.999	9.129	13	25	38	50	63	76	88	101	113
2.1	9.261	9.394	9.528	9.664	9.800	9.938					14	27	41	54	69	81	95	108	122
2.1							10.08	10.22	10.36	10.50	1	3	4	6	7	9	10	11	13
2.2	10.65	10.79	10.94	11.09	11.24	11.39	11.54	11.70	11.85	12.01	2	3	5	6	8	9	11	12	14
2.3	12.17	12.33	12.49	12.65	12.81	12.98	13.14	13.31	13.48	13.65	2	3	5	7	8	10	12	13	15
2.4	13.82	14.00	14.17	14.35	14.53	14.71	14.89	15.07	15.25	15.44	2	4	5	7	9	11	13	14	16
2.5	15.63	15.81	16.00	16.19	16.39	16.58	16.78	16.97	17.17	17.37	2	4	6	8	10	12	14	16	18
2.6	17.58	17.78	17.98	18.19	18.40	18.61	18.82	19.03	19.25	19.47	2	4	6	8	11	13	15	17	19
2.7	19.68	19.90	20.12	20.35	20.57	20.80	21.02	21.25	21.48	21.72	2	5	7	9	11	14	16	18	20
2.8	21.95	22.19	22.43	22.67	22.91	23.15	23.39	23.64	23.89	24.14	2	5	7	10	12	15	17	19	22
2.9	24.39	24.64	24.90	25.15	25.41	25.67	25.93	26.20	26.46	26.73	3	5	8	10	13	16	18	21	23
3.0	27.00	27.27	27.54	27.82	28.09	28.37	28.65	28.93	29.22	29.50	3	6	8	11	14	17	20	22	25
3.1	29.79	30.08	30.37	30.66	30.96	31.26	31.55	31.86	32.16	32.46	3	6	9	12	15	18	21	24	27
3.2	32.77	33.08	33.39	33.70	34.01	34.33	34.65	34.97	35.29	35.61	3	6	9	13	16	19	22	25	29
3.3	35.94	36.26	36.59	36.93	37.26	37.60	37.93	38.27	38.61	38.96	3	7	10	13	17	20	24	27	30
3.4	39.30	39.65	40.00	40.35	40.71	41.06	41.42	41.78	42.14	42.51	4	7	11	14	18	21	25	29	32
3.5	42.88	43.24	43.61	43.99	44.36	44.74	45.12	45.50	45.88	46.27	4	8	11	15	19	23	26	30	34
3.6	46.66	47.05	47.44	47.83	48.23	48.63	49.03	49.43	49.84	50.24	4	8	12	16	20	24	28	32	36
3.7	50.65	51.06	51.48	51.90	52.31	52.73	53.16	53.58	54.01	54.44	4	8	13	17	21	25	30	34	38
3.8	54.87	55.31	55.74	56.18	56.62	57.07	57.51	57.96	58.41	58.86	4	9	13	18	22	27	31	36	40
3.9	59.32	59.78	60.24	60.70	61.16	61.63	62.10	62.57	63.04	63.52	5	9	14	19	23	28	33	37	42
4.0	64.00	64.48	64.96	65.45	65.94	66.43	66.92	67.42	67.92	68.42	5	10	15	20	25	29	34	39	44
4.1	68.92	69.43	69.93	70.44	70.96	71.47	71.99	72.51	73.03	73.56	5	10	15	21	26	31	36	41	46
4.2	74.09	74.62	75.15	75.69	76.23	76.77	77.31	77.85	78.40	78.95	5	11	16	22	27	32	38	43	49
4.3	79.51	80.06	80.62	81.18	81.75	82.31	82.88	83.45	84.03	84.60	6	11	17	23	28	34	40	45	51
4.4	85.18	85.77	86.35	86.94	87.53	88.12	88.72	89.31	89.92	90.52	6	12	18	24	30	36	42	48	53
4.5	91.13	91.73	92.35	92.96	93.58	94.20	94.82	95.44	96.07	96.70	6	12	19	25	31	37	43	50	56
4.6	97.34	97.97	98.61	99.25	99.90						6	13	19	26	32	38	45	51	58
4.6						100.5	101.2	101.8	102.5	103.2	1	1	2	3	3	4	5	5	6
4.7	103.8	104.5	105.2	105.8	106.5	107.2	107.9	108.5	109.2	109.9	1	1	2	3	3	4	5	5	6
4.8	110.6	111.3	112.0	112.7	113.4	114.1	114.8	115.5	116.2	116.9	1	1	2	3	4	4	5	6	6
4.9	117.6	118.4	119.1	119.8	120.6	121.3	122.0	122.8	123.5	124.3	1	1	2	3	4	4	5	6	7
5.0	125.0	125.8	126.5	127.3	128.0	128.8	129.6	130.3	131.1	131.9	1	2	2	3	4	5	5	6	7
5.1	132.7	133.4	134.2	135.0	135.8	136.6	137.4	138.2	139.0	139.8	1	2	2	3	4	5	6	6	7
5.2	140.6	141.4	142.2	143.1	143.9	144.7	145.5	146.4	147.2	148.0	1	2	2	3	4	5	6	7	7
5.3	148.9	149.7	150.6	151.4	152.3	153.1	154.0	154.9	155.7	156.6	1	2	3	3	4	5	6	7	8
5.4	157.5	158.3	159.2	160.1	161.0	161.9	162.8	163.7	164.6	165.5	1	2	3	4	4	5	6	7	8

$$5.625^3 = 178.0$$
$$562.5^3 = (5,625 \times 100)^3 = 178.0 \times 100^3 = 178\,000\,000$$
$$0.5625^3 = (5.625 \times \tfrac{1}{10})^3 = 178.0 \times \tfrac{1}{1000} = 0.178$$

Cubes

x	0	1	2	3	4	5	6	7	8	9	Add 1	2	3	4	5	6	7	8	9
5.5	166.4	167.3	168.2	169.1	170.0	171.0	171.9	172.8	173.7	174.7	1	2	3	4	5	6	6	7	8
5.6	175.6	176.6	177.5	178.5	179.4	180.4	181.3	182.3	183.3	184.2	1	2	3	4	5	6	7	8	9
5.7	185.2	186.2	187.1	188.1	189.1	190.1	191.1	192.1	193.1	194.1	1	2	3	4	5	6	7	8	9
5.8	195.1	196.1	197.1	198.2	199.2	200.2	201.2	202.3	203.3	204.3	1	2	3	4	5	6	7	8	9
5.9	205.4	206.4	207.5	208.5	209.6	210.6	211.7	212.8	213.8	214.9	1	2	3	4	5	6	7	8	10
6.0	216.0	217.1	218.2	219.3	220.3	221.4	222.5	223.6	224.8	225.9	1	2	3	4	5	7	8	9	10
6.1	227.0	228.1	229.2	230.3	231.5	232.6	233.7	234.9	236.0	237.2	1	2	3	5	6	7	8	9	10
6.2	238.3	239.5	240.6	241.8	243.0	244.1	245.3	246.5	247.7	248.9	1	2	4	5	6	7	8	9	11
6.3	250.0	251.2	252.4	253.6	254.8	256.0	257.3	258.5	259.7	260.9	1	2	4	5	6	7	8	10	11
6.4	262.1	263.4	264.6	265.8	267.1	268.3	269.6	270.8	272.1	273.4	1	2	4	5	6	7	9	10	11
6.5	274.6	275.9	277.2	278.4	279.7	281.0	282.3	283.6	284.9	286.2	1	3	4	5	6	8	9	10	12
6.6	287.5	288.8	290.1	291.4	292.8	294.1	295.4	296.7	298.1	299.4	1	3	4	5	7	8	9	11	12
6.7	300.8	302.1	303.5	304.8	306.2	307.5	308.9	310.3	311.7	313.0	1	3	4	5	7	8	10	11	12
6.8	314.4	315.8	317.2	318.6	320.0	321.4	322.8	324.2	325.7	327.1	1	3	4	6	7	8	10	11	13
6.9	328.5	329.9	331.4	332.8	334.3	335.7	337.2	338.6	340.1	341.5	1	3	4	6	7	9	10	12	13
7.0	343.0	344.5	345.9	347.4	348.9	350.4	351.9	353.4	354.9	356.4	1	3	4	6	7	9	10	12	13
7.1	357.9	359.4	360.9	362.5	364.0	365.5	367.1	368.6	370.1	371.7	2	3	5	6	8	9	11	12	14
7.2	373.2	374.8	376.4	377.9	379.5	381.1	382.7	384.2	385.8	387.4	2	3	5	6	8	9	11	13	14
7.3	389.0	390.6	392.2	393.8	395.4	397.1	398.7	400.3	401.9	403.6	2	3	5	6	8	10	11	13	15
7.4	405.2	406.9	408.5	410.2	411.8	413.5	415.2	416.8	418.5	420.2	2	3	5	7	8	10	12	13	15
7.5	421.9	423.6	425.3	427.0	428.7	430.4	432.1	433.8	435.5	437.2	2	3	5	7	9	10	12	14	15
7.6	439.0	440.7	442.5	444.2	445.9	447.7	449.5	451.2	453.0	454.8	2	4	5	7	9	11	12	14	16
7.7	456.5	458.3	460.1	461.9	463.7	465.5	467.3	469.1	470.9	472.7	2	4	5	7	9	11	13	14	16
7.8	474.6	476.4	478.2	480.0	481.9	483.7	485.6	487.4	489.3	491.2	2	4	6	7	9	11	13	15	17
7.9	493.0	494.9	496.8	498.7	500.6	502.5	504.4	506.3	508.2	510.1	2	4	6	8	9	11	13	15	17
8.0	512.0	513.9	515.8	517.8	519.7	521.7	523.6	525.6	527.5	529.5	2	4	6	8	10	12	14	16	17
8.1	531.4	533.4	535.4	537.4	539.4	541.3	543.3	545.3	547.3	549.4	2	4	6	8	10	12	14	16	18
8.2	551.4	553.4	555.4	557.4	559.5	561.5	563.6	565.6	567.7	569.7	2	4	6	8	10	12	14	16	18
8.3	571.8	573.9	575.9	578.0	580.1	582.2	584.3	586.4	588.5	590.6	2	4	6	8	10	13	15	17	19
8.4	592.7	594.8	596.9	599.1	601.2	603.4	605.5	607.6	609.8	611.9	2	4	6	9	11	13	15	17	19
8.5	614.1	616.3	618.5	620.7	622.8	625.0	627.2	629.4	631.6	633.8	2	4	7	9	11	13	15	18	20
8.6	636.1	638.3	640.5	642.7	645.0	647.2	649.5	651.7	654.0	656.2	2	4	7	9	11	13	16	18	20
8.7	658.5	660.8	663.1	665.3	667.6	669.9	672.2	674.5	676.8	679.2	2	5	7	9	11	14	16	18	21
8.8	681.5	683.8	686.1	688.5	690.8	693.2	695.5	697.9	700.2	702.6	2	5	7	9	12	14	16	19	21
8.9	705.0	707.3	709.7	712.1	714.5	716.9	719.3	721.7	724.2	726.6	2	5	7	10	12	14	17	19	22
9.0	729.0	731.4	733.9	736.3	738.8	741.2	743.7	746.1	748.6	751.1	3	5	7	10	12	15	17	20	22
9.1	753.6	756.1	758.6	761.0	763.6	766.1	768.6	771.1	773.6	776.2	3	5	8	10	13	15	18	20	23
9.2	778.7	781.2	783.8	786.3	788.9	791.5	794.0	796.6	799.2	801.8	3	5	8	10	13	15	18	21	23
9.3	804.4	807.0	809.6	812.2	814.8	817.4	820.0	822.7	825.3	827.9	3	5	8	10	13	16	18	21	24
9.4	830.6	833.2	835.9	838.6	841.2	843.9	846.6	849.3	852.0	854.7	3	5	8	11	13	16	19	21	24
9.5	857.4	860.1	862.8	865.5	868.3	871.0	873.7	876.5	879.2	882.0	3	5	8	11	14	16	19	22	25
9.6	884.7	887.5	890.3	893.1	895.8	898.6	901.4	904.2	907.0	909.9	3	6	8	11	14	17	20	22	25
9.7	912.7	915.5	918.3	921.2	924.0	926.9	929.7	932.6	935.4	938.3	3	6	9	11	14	17	20	23	26
9.8	941.2	944.1	947.0	949.9	952.8	955.7	958.6	961.5	964.4	967.4	3	6	9	12	15	17	20	23	26
9.9	970.3	973.2	976.2	979.1	982.1	985.1	988.0	991.0	994.0	997.0	3	6	9	12	15	18	21	24	27

$$\sqrt[3]{1.875} = 1.233, \quad \sqrt[3]{18.75} = 2.656, \quad \sqrt[3]{187.5} = 5.724$$

$$\sqrt[3]{18.750} = \sqrt[3]{18.75 \times 1000} = 2.656 \times 10 = 26.56$$

$$\sqrt[3]{0.001875} = \sqrt[3]{1.875 \times \tfrac{1}{1000}} = 1.233 \times \tfrac{1}{10} = 0.1233$$

Table of Reciprocals of Numbers from 1–10

	0	1	2	3	4	5	6	7	8	9	1	2	3	4	5	6	7	8	9
1.0	1.0000	0.9901	0.9804	0.9709	0.9615	0.9524	0.9434	0.9346	0.9259	0.9174									
1.1	0.9091	0.9009	0.8929	0.8850	0.8772	0.8696	0.8621	0.8547	0.8475	0.8403									
1.2	0.8333	0.8264	0.8197	0.8130	0.8065	0.8000	0.7937	0.7874	0.7813	0.7752									
1.3	0.7692	0.7634	0.7576	0.7519	0.7463	0.7407	0.7353	0.7299	0.7246	0.7194									
1.4	0.7143	0.7092	0.7042	0.6993	0.6944	0.6897	0.6849	0.6803	0.6757	0.6711									
1.5	0.6667	0.6623	0.6579	0.6536	0.6494	0.6452	0.6410	0.6369	0.6329	0.6289	4	8	12	17	21	25	29	33	37
1.6	0.6250	0.6211	0.6173	0.6135	0.6098	0.6061	0.6024	0.5988	0.5952	0.5917	4	7	11	15	18	22	26	29	33
1.7	0.5882	0.5848	0.5814	0.5780	0.5747	0.5714	0.5682	0.5650	0.5618	0.5587	3	7	10	13	16	20	23	26	29
1.8	0.5556	0.5525	0.5495	0.5464	0.5435	0.5405	0.5376	0.5348	0.5319	0.5291	3	6	9	12	15	18	20	23	26
1.9	0.5263	0.5236	0.5208	0.5181	0.5155	0.5128	0.5102	0.5076	0.5051	0.5025	3	5	8	11	13	16	18	21	24
2.0	0.5000	0.4975	0.4950	0.4926	0.4902	0.4878	0.4854	0.4831	0.4808	0.4785	2	5	7	10	12	14	17	19	21
2.1	0.4762	0.4739	0.4717	0.4695	0.4673	0.4651	0.4630	0.4608	0.4587	0.4566	2	4	6	9	11	13	15	17	19
2.2	0.4545	0.4525	0.4505	0.4484	0.4464	0.4444	0.4425	0.4405	0.4386	0.4367	2	4	6	8	10	12	14	16	18
2.3	0.4348	0.4329	0.4310	0.4292	0.4274	0.4255	0.4237	0.4219	0.4202	0.4184	2	4	5	7	9	11	13	14	16
2.4	0.4167	0.4149	0.4132	0.4115	0.4098	0.4082	0.4065	0.4049	0.4032	0.4016	2	3	5	7	8	10	12	13	15
2.5	0.4000	0.3984	0.3968	0.3953	0.3937	0.3922	0.3906	0.3891	0.3876	0.3861	2	3	5	6	8	9	11	12	14
2.6	0.3846	0.3831	0.3817	0.3802	0.3788	0.3774	0.3759	0.3745	0.3731	0.3717	1	3	4	6	7	9	10	11	13
2.7	0.3704	0.3690	0.3676	0.3663	0.3650	0.3636	0.3623	0.3610	0.3597	0.3584	1	3	4	5	7	8	9	11	12
2.8	0.3571	0.3559	0.3546	0.3534	0.3521	0.3509	0.3497	0.3484	0.3472	0.3460	1	2	4	5	6	7	9	10	11
2.9	0.3448	0.3436	0.3425	0.3413	0.3401	0.3390	0.3378	0.3367	0.3356	0.3344	1	2	3	5	6	7	8	9	10
3.0	0.3333	0.3322	0.3311	0.3300	0.3289	0.3279	0.3268	0.3257	0.3247	0.3236	1	2	3	4	5	6	8	9	10
3.1	0.3226	0.3215	0.3205	0.3195	0.3185	0.3175	0.3165	0.3155	0.3145	0.3135	1	2	3	4	5	6	7	8	9
3.2	0.3125	0.3115	0.3106	0.3096	0.3086	0.3077	0.3067	0.3058	0.3049	0.3040	1	2	3	4	5	6	7	8	9
3.3	0.3030	0.3021	0.3012	0.3003	0.2994	0.2985	0.2976	0.2967	0.2959	0.2950	1	2	3	4	5	6	7	8	
3.4	0.2941	0.2933	0.2924	0.2915	0.2907	0.2899	0.2890	0.2882	0.2874	0.2865	1	2	3	4	5	6	7	8	
3.5	0.2857	0.2849	0.2841	0.2833	0.2825	0.2817	0.2809	0.2801	0.2793	0.2786	1	2	2	3	4	5	6	6	7
3.6	0.2778	0.2770	0.2762	0.2755	0.2747	0.2740	0.2732	0.2725	0.2717	0.2710	1	2	2	3	4	5	5	6	7
3.7	0.2703	0.2695	0.2688	0.2681	0.2674	0.2667	0.2660	0.2653	0.2646	0.2639	1	1	2	3	4	4	5	6	6
3.8	0.2632	0.2625	0.2618	0.2611	0.2604	0.2597	0.2591	0.2584	0.2577	0.2571	1	1	2	3	3	4	5	5	6
3.9	0.2564	0.2558	0.2551	0.2545	0.2538	0.2532	0.2525	0.2519	0.2513	0.2506	1	1	2	3	3	4	4	5	6
4.0	0.2500	0.2494	0.2488	0.2481	0.2475	0.2469	0.2463	0.2457	0.2451	0.2445	1	1	2	2	3	4	4	5	5
4.1	0.2439	0.2433	0.2427	0.2421	0.2415	0.2410	0.2404	0.2398	0.2392	0.2387	1	1	2	2	3	4	4	5	5
4.2	0.2381	0.2375	0.2370	0.2364	0.2358	0.2353	0.2347	0.2342	0.2336	0.2331	1	1	2	2	3	3	4	4	5
4.3	0.2326	0.2320	0.2315	0.2309	0.2304	0.2299	0.2294	0.2288	0.2283	0.2278	1	1	2	2	3	3	4	4	5
4.4	0.2273	0.2268	0.2262	0.2257	0.2252	0.2247	0.2242	0.2237	0.2232	0.2227	1	1	2	2	3	3	4	4	5
4.5	0.2222	0.2217	0.2212	0.2208	0.2203	0.2198	0.2193	0.2188	0.2183	0.2179	0	1	1	2	2	3	3	4	4
4.6	0.2174	0.2169	0.2165	0.2160	0.2155	0.2151	0.2146	0.2141	0.2137	0.2132	0	1	1	2	2	3	3	4	4
4.7	0.2128	0.2123	0.2119	0.2114	0.2110	0.2105	0.2101	0.2096	0.2092	0.2088	0	1	1	2	2	3	3	4	4
4.8	0.2083	0.2079	0.2075	0.2070	0.2066	0.2062	0.2058	0.2053	0.2049	0.2045	0	1	1	2	2	3	3	3	4
4.9	0.2041	0.2037	0.2033	0.2028	0.2024	0.2020	0.2016	0.2012	0.2008	0.2004	0	1	1	2	2	3	3	3	4
5.0	0.2000	0.1996	0.1992	0.1988	0.1984	0.1980	0.1976	0.1972	0.1969	0.1965	0	1	1	2	2	3	3	3	4
5.1	0.1961	0.1957	0.1953	0.1949	0.1946	0.1942	0.1938	0.1934	0.1931	0.1927	0	1	1	2	2	2	3	3	3
5.2	0.1923	0.1919	0.1916	0.1912	0.1908	0.1905	0.1901	0.1898	0.1894	0.1890	0	1	1	1	2	2	3	3	3
5.3	0.1887	0.1883	0.1880	0.1876	0.1873	0.1869	0.1866	0.1862	0.1859	0.1855	0	1	1	1	2	2	3	3	3
5.4	0.1852	0.1848	0.1845	0.1842	0.1838	0.1835	0.1832	0.1828	0.1825	0.1821	0	1	1	1	2	2	2	3	3

$$\text{Reciprocal of a number} = \frac{1}{\text{number}}$$

$$\text{Reciprocal of } 2.361 = \frac{1}{2.361} = 0.4235$$

To find reciprocals of numbers outside the range of 1 to 10 proceed as shown on page 79.

Table of Reciprocals of Numbers from 1–10

	0	1	2	3	4	5	6	7	8	9	1	2	3	4	5	6	7	8	9
5.5	0.1818	0.1815	0.1812	0.1808	0.1805	0.1802	0.1799	0.1795	0.1792	0.1789	0	1	1	1	2	2	2	3	3
5.6	0.1786	0.1783	0.1779	0.1776	0.1773	0.1770	0.1767	0.1764	0.1761	0.1757	0	1	1	1	2	2	2	3	3
5.7	0.1754	0.1751	0.1748	0.1745	0.1742	0.1739	0.1736	0.1733	0.1730	0.1727	0	1	1	1	2	2	2	2	3
5.8	0.1724	0.1721	0.1718	0.1715	0.1712	0.1709	0.1706	0.1704	0.1701	0.1698	0	1	1	1	1	2	2	2	3
5.9	0.1695	0.1692	0.1689	0.1686	0.1684	0.1681	0.1678	0.1675	0.1672	0.1669	0	1	1	1	1	2	2	2	3
6.0	0.1667	0.1664	0.1661	0.1658	0.1656	0.1653	0.1650	0.1647	0.1645	0.1642	0	1	1	1	1	2	2	2	2
6.1	0.1639	0.1637	0.1634	0.1631	0.1629	0.1626	0.1623	0.1621	0.1618	0.1616	0	1	1	1	1	2	2	2	2
6.2	0.1613	0.1610	0.1608	0.1605	0.1603	0.1600	0.1597	0.1595	0.1592	0.1590	0	1	1	1	1	2	2	2	2
6.3	0.1587	0.1585	0.1582	0.1580	0.1577	0.1575	0.1572	0.1570	0.1567	0.1565	0	0	1	1	1	1	2	2	2
6.4	0.1563	0.1560	0.1558	0.1555	0.1553	0.1550	0.1548	0.1546	0.1543	0.1541	0	0	1	1	1	1	2	2	2
6.5	0.1538	0.1536	0.1534	0.1531	0.1529	0.1527	0.1524	0.1522	0.1520	0.1517	0	0	1	1	1	1	2	2	2
6.6	0.1515	0.1513	0.1511	0.1508	0.1506	0.1504	0.1502	0.1499	0.1497	0.1495	0	0	1	1	1	1	2	2	2
6.7	0.1493	0.1490	0.1488	0.1486	0.1484	0.1481	0.1479	0.1477	0.1475	0.1473	0	0	1	1	1	1	2	2	2
6.8	0.1471	0.1468	0.1466	0.1464	0.1462	0.1460	0.1458	0.1456	0.1453	0.1451	0	0	1	1	1	1	1	2	2
6.9	0.1449	0.1447	0.1445	0.1443	0.1441	0.1439	0.1437	0.1435	0.1433	0.1431	0	0	1	1	1	1	1	2	2
7.0	0.1429	0.1427	0.1425	0.1422	0.1420	0.1418	0.1416	0.1414	0.1412	0.1410	0	0	1	1	1	1	1	2	2
7.1	0.1408	0.1406	0.1404	0.1403	0.1401	0.1399	0.1397	0.1395	0.1393	0.1391	0	0	1	1	1	1	1	2	2
7.2	0.1389	0.1387	0.1385	0.1383	0.1381	0.1379	0.1377	0.1376	0.1374	0.1372	0	0	1	1	1	1	1	2	2
7.3	0.1370	0.1368	0.1366	0.1364	0.1362	0.1361	0.1359	0.1357	0.1355	0.1353	0	0	1	1	1	1	1	1	2
7.4	0.1351	0.1350	0.1348	0.1346	0.1344	0.1342	0.1340	0.1339	0.1337	0.1335	0	0	1	1	1	1	1	1	2
7.5	0.1333	0.1332	0.1330	0.1328	0.1326	0.1325	0.1323	0.1321	0.1319	0.1318	0	0	1	1	1	1	1	1	2
7.6	0.1316	0.1314	0.1312	0.1311	0.1309	0.1307	0.1305	0.1304	0.1302	0.1300	0	0	1	1	1	1	1	1	1
7.7	0.1299	0.1297	0.1295	0.1294	0.1292	0.1290	0.1289	0.1287	0.1285	0.1284	0	0	0	1	1	1	1	1	1
7.8	0.1282	0.1280	0.1279	0.1277	0.1276	0.1274	0.1272	0.1271	0.1269	0.1267	0	0	0	1	1	1	1	1	1
7.9	0.1266	0.1264	0.1263	0.1261	0.1259	0.1258	0.1256	0.1255	0.1253	0.1252	0	0	0	1	1	1	1	1	1
8.0	0.1250	0.1248	0.1247	0.1245	0.1244	0.1242	0.1241	0.1239	0.1238	0.1236	0	0	0	1	1	1	1	1	1
8.1	0.1235	0.1233	0.1232	0.1230	0.1229	0.1227	0.1225	0.1224	0.1222	0.1221	0	0	0	1	1	1	1	1	1
8.2	0.1220	0.1218	0.1217	0.1215	0.1214	0.1212	0.1211	0.1209	0.1208	0.1206	0	0	0	1	1	1	1	1	1
8.3	0.1205	0.1203	0.1202	0.1200	0.1199	0.1198	0.1196	0.1195	0.1193	0.1192	0	0	0	1	1	1	1	1	1
8.4	0.1190	0.1189	0.1188	0.1186	0.1185	0.1183	0.1182	0.1181	0.1179	0.1178	0	0	0	1	1	1	1	1	1
8.5	0.1176	0.1175	0.1174	0.1172	0.1171	0.1170	0.1168	0.1167	0.1166	0.1164	0	0	0	1	1	1	1	1	1
8.6	0.1163	0.1161	0.1160	0.1159	0.1157	0.1156	0.1155	0.1153	0.1152	0.1151	0	0	0	1	1	1	1	1	1
8.7	0.1149	0.1148	0.1147	0.1145	0.1144	0.1143	0.1142	0.1140	0.1139	0.1138	0	0	0	1	1	1	1	1	1
8.8	0.1136	0.1135	0.1134	0.1133	0.1131	0.1130	0.1129	0.1127	0.1126	0.1125	0	0	0	1	1	1	1	1	1
8.9	0.1124	0.1122	0.1121	0.1120	0.1119	0.1117	0.1116	0.1115	0.1114	0.1112	0	0	0	0	1	1	1	1	1
9.0	0.1111	0.1110	0.1109	0.1107	0.1106	0.1105	0.1104	0.1103	0.1101	0.1100	0	0	0	0	1	1	1	1	1
9.1	0.1099	0.1098	0.1096	0.1095	0.1094	0.1093	0.1092	0.1091	0.1089	0.1088	0	0	0	0	1	1	1	1	1
9.2	0.1087	0.1086	0.1085	0.1083	0.1082	0.1081	0.1080	0.1079	0.1078	0.1076	0	0	0	0	1	1	1	1	1
9.3	0.1075	0.1074	0.1073	0.1072	0.1071	0.1070	0.1068	0.1067	0.1066	0.1065	0	0	0	0	1	1	1	1	1
9.4	0.1064	0.1063	0.1062	0.1060	0.1059	0.1058	0.1057	0.1056	0.1055	0.1054	0	0	0	0	1	1	1	1	1
9.5	0.1053	0.1052	0.1050	0.1049	0.1048	0.1047	0.1046	0.1045	0.1044	0.1043	0	0	0	0	1	1	1	1	1
9.6	0.1042	0.1041	0.1040	0.1038	0.1037	0.1036	0.1035	0.1034	0.1033	0.1032	0	0	0	0	1	1	1	1	1
9.7	0.1031	0.1030	0.1029	0.1028	0.1027	0.1026	0.1025	0.1024	0.1022	0.1021	0	0	0	0	1	1	1	1	1
9.8	0.1020	0.1019	0.1018	0.1017	0.1016	0.1015	0.1014	0.1013	0.1012	0.1011	0	0	0	0	1	1	1	1	1
9.9	0.1010	0.1009	0.1008	0.1007	0.1006	0.1005	0.1004	0.1003	0.1002	0.1001	0	0	0	0	1	1	1	1	1

To find $\dfrac{1}{639.2}$

$$\frac{1}{639.2} = \frac{1}{6.392} \times \frac{1}{100} = 0.1565 \times \frac{1}{100} = 0.1565 \times 10^{-2}$$
$$\text{or } 0.001\,565$$

To find $\dfrac{1}{0.039\,82}$

$$\frac{1}{0.039\,82} = \frac{1}{3.982} \times \frac{100}{1} = 0.2512 \times 10^{2} = 25.12$$

Natural Logarithms

	0	1	2	3	4	5	6	7	8	9	1	2	3	4	5	6	7	8	9
1.0	0.0000	0.0100	0.0198	0.0296	0.0392	0.0488	0.0583	0.0677	0.0770	0.0862	10	19	29	38	48	57	67	76	86
1.1	0.0953	0.1044	0.1133	0.1222	0.1310	0.1398	0.1484	0.1570	0.1655	0.1740	9	17	26	35	44	52	61	70	78
1.2	0.1823	0.1906	0.1989	0.2070	0.2151	0.2231	0.2311	0.2390	0.2469	0.2546	8	16	24	32	40	48	56	64	72
1.3	0.2624	0.2700	0.2776	0.2852	0.2927	0.3001	0.3075	0.3148	0.3221	0.3293	7	15	22	30	37	44	52	59	67
1.4	0.3365	0.3436	0.3507	0.3577	0.3646	0.3716	0.3784	0.3853	0.3920	0.3988	7	14	21	28	34	41	48	55	62
1.5	0.4055	0.4121	0.4187	0.4253	0.4318	0.4383	0.4447	0.4511	0.4574	0.4637	6	13	19	26	32	39	45	52	58
1.6	0.4700	0.4762	0.4824	0.4886	0.4947	0.5008	0.5068	0.5128	0.5188	0.5247	6	12	18	24	30	36	42	48	55
1.7	0.5306	0.5365	0.5423	0.5481	0.5539	0.5596	0.5653	0.5710	0.5766	0.5822	6	11	17	23	29	34	40	46	51
1.8	0.5878	0.5933	0.5988	0.6043	0.6098	0.6152	0.6206	0.6259	0.6313	0.6366	5	11	16	22	27	32	38	43	49
1.9	0.6419	0.6471	0.6523	0.6575	0.6627	0.6678	0.6729	0.6780	0.6831	0.6881	5	10	15	21	26	31	36	41	46
2.0	0.6931	0.6981	0.7031	0.7080	0.7129	0.7178	0.7227	0.7275	0.7324	0.7372	5	10	15	20	24	29	34	39	44
2.1	0.7419	0.7467	0.7514	0.7561	0.7608	0.7655	0.7701	0.7747	0.7793	0.7839	5	9	14	19	23	28	33	37	42
2.2	0.7885	0.7930	0.7975	0.8020	0.8065	0.8109	0.8154	0.8198	0.8242	0.8286	4	9	13	18	22	27	31	36	40
2.3	0.8329	0.8372	0.8416	0.8459	0.8502	0.8544	0.8587	0.8629	0.8671	0.8713	4	9	13	17	21	26	30	34	38
2.4	0.8755	0.8796	0.8838	0.8879	0.8920	0.8961	0.9002	0.9042	0.9083	0.9123	4	8	12	16	20	24	29	33	37
2.5	0.9163	0.9203	0.9243	0.9282	0.9322	0.9361	0.9400	0.9349	0.9478	0.9517	4	8	12	16	20	24	27	31	35
2.6	0.9555	0.9594	0.9632	0.9670	0.9708	0.9746	0.9783	0.9821	0.9858	0.9895	4	8	11	15	19	23	26	30	34
2.7	0.9933	0.9969	1.0006	1.0043	1.0080	1.0116	1.0152	1.0188	1.0225	1.0260	4	7	11	15	18	22	25	29	33
2.8	1.0269	1.0332	1.0367	1.0403	1.0438	1.0473	1.0508	1.0543	1.0578	1.0613	4	7	11	14	18	21	25	28	32
2.9	1.0647	1.0682	1.0716	1.0750	1.0784	1.0818	1.0852	1.0886	1.0919	1.0953	3	7	10	14	17	20	24	27	31
3.0	1.0986	1.1019	1.1053	1.1086	1.1119	1.1151	1.1184	1.1217	1.1249	1.1282	3	7	10	13	16	20	23	26	30
3.1	1.1314	1.1346	1.1378	1.1410	1.1442	1.1474	1.1506	1.1537	1.1569	1.1600	3	6	10	13	16	19	22	25	29
3.2	1.1632	1.1663	1.1694	1.1725	1.1756	1.1787	1.1817	1.1848	1.1878	1.1909	3	6	9	12	15	18	22	25	28
3.3	1.1939	1.1969	1.2000	1.2030	1.2060	1.2090	1.2119	1.2149	1.2179	1.2208	3	6	9	12	15	18	21	24	27
3.4	1.2238	1.2267	1.2296	1.2326	1.2355	1.2384	1.2413	1.2442	1.2470	1.2499	3	6	9	12	14	17	20	23	26
3.5	1.2528	1.2556	1.2585	1.2613	1.2641	1.2669	1.2698	1.2726	1.2754	1.2782	3	6	8	11	14	17	20	23	25
3.6	1.2809	1.2837	1.2865	1.2892	1.2920	1.2947	1.2975	1.3002	1.3029	1.3056	3	5	8	11	14	16	19	22	25
3.7	1.3083	1.3110	1.3137	1.3164	1.3191	1.3218	1.3244	1.3271	1.3297	1.3324	3	5	8	11	13	16	19	21	24
3.8	1.3350	1.3376	1.3403	1.3429	1.3455	1.3481	1.3507	1.3533	1.3558	1.3584	3	5	8	10	13	16	18	21	23
3.9	1.3610	1.3635	1.3661	1.3686	1.3712	1.3737	1.3762	1.3788	1.3813	1.3838	3	5	8	10	13	15	18	20	23
4.0	1.3863	1.3888	1.3913	1.3938	1.3962	1.3987	1.4012	1.4036	1.4061	1.4085	2	5	7	10	12	15	17	20	22
4.1	1.4110	1.4134	1.4159	1.4183	1.4207	1.4231	1.4255	1.4279	1.4303	1.4327	2	5	7	10	12	14	17	19	22
4.2	1.4351	1.4375	1.4398	1.4422	1.4446	1.4469	1.4493	1.4516	1.4540	1.4563	2	5	7	9	12	14	16	19	21
4.3	1.4586	1.4609	1.4633	1.4656	1.4679	1.4702	1.4725	1.4748	1.4770	1.4793	2	5	7	9	11	14	16	18	21
4.4	1.4816	1.4839	1.4861	1.4884	1.4907	1.4929	1.4951	1.4974	1.4996	1.5019	2	4	7	9	11	13	16	18	20
4.5	1.5041	1.5063	1.5085	1.5107	1.5129	1.5151	1.5173	1.5195	1.5217	1.5239	2	4	7	9	11	13	15	18	20
4.6	1.5261	1.5282	1.5304	1.5326	1.5347	1.5369	1.5390	1.5412	1.5433	1.5454	2	4	6	9	11	13	15	17	19
4.7	1.5476	1.5497	1.5518	1.5539	1.5560	1.5581	1.5602	1.5623	1.5644	1.5665	2	4	6	8	11	13	15	17	19
4.8	1.5686	1.5707	1.5728	1.5748	1.5769	1.5790	1.5810	1.5831	1.5851	1.5877	2	4	6	8	10	12	14	16	19
4.9	1.5892	1.5913	1.5933	1.5953	1.5974	1.5994	1.6014	1.6034	1.6054	1.6074	2	4	6	8	10	12	14	16	18
5.0	1.6094	1.6114	1.6134	1.6154	1.6174	1.6194	1.6214	1.6233	1.6253	1.6273	2	4	6	8	10	12	14	16	18
5.1	1.6292	1.6312	1.6332	1.6351	1.6371	1.6390	1.6409	1.6429	1.6448	1.6467	2	4	6	8	10	12	14	16	17
5.2	1.6487	1.6506	1.6525	1.6544	1.6563	1.6582	1.6601	1.6620	1.6639	1.6658	2	4	6	8	10	11	13	15	17
5.3	1.6677	1.6696	1.6715	1.6734	1.6752	1.6771	1.6790	1.6808	1.6827	1.6845	2	4	6	7	9	11	13	15	17
5.4	1.6864	1.6882	1.6901	1.6919	1.6938	1.6956	1.6974	1.6993	1.7011	1.7029	2	4	6	7	9	11	13	15	17

Natural Logarithms of 10^{+n}

n	1	2	3	4	5	6	7	8	9
$\ln 10^{+n}$	2.3026	4.6052	6.9078	9.2103	11.5129	13.8155	16.1181	18.4207	20.7233

To find $\log_e 483.4$

$$483.4 = 4.834 \times 100 = 4.834 \times 10^2$$
$$\log_e 483.4 = \log_e 4.834 + \log_e 10^2$$
$$= 1.5756 + 4.6052 = 6.1808$$

Natural Logarithms

	0	1	2	3	4	5	6	7	8	9	1	2	3	4	5	6	7	8	9
5.5	1.7047	1.7066	1.7084	1.7102	1.7120	1.7138	1.7156	1.7174	1.7192	1.7210	2	4	5	7	9	11	13	14	16
5.6	1.7228	1.7246	1.7263	1.7281	1.7299	1.7317	1.7334	1.7352	1.7370	1.7387	2	4	5	7	9	11	12	14	16
5.7	1.7405	1.7422	1.7440	1.7457	1.7475	1.7492	1.7509	1.7527	1.7544	1.7561	2	3	5	7	9	10	12	14	16
5.8	1.7579	1.7596	1.7613	1.7630	1.7647	1.7664	1.7681	1.7699	1.7716	1.7733	2	3	5	7	9	10	12	14	15
5.9	1.7750	1.7766	1.7783	1.7800	1.7817	1.7834	1.7851	1.7867	1.7884	1.7901	2	3	5	7	8	10	12	13	15
6.0	1.7918	1.7934	1.7951	1.7967	1.7984	1.8001	1.8017	1.8034	1.8050	1.8066	2	3	5	7	8	10	12	13	15
6.1	1.8083	1.8099	1.8116	1.8132	1.8148	1.8165	1.8181	1.8197	1.8213	1.8229	2	3	5	7	8	10	11	13	15
6.2	1.8245	1.8262	1.8278	1.8294	1.8310	1.8326	1.8342	1.8358	1.8374	1.8390	2	3	5	6	8	10	11	13	14
6.3	1.8405	1.8421	1.8437	1.8453	1.8469	1.8485	1.8500	1.8516	1.8532	1.8547	2	3	5	6	8	9	11	13	14
6.4	1.8563	1.8579	1.8594	1.8610	1.8625	1.8641	1.8656	1.8672	1.8687	1.8708	2	3	5	6	8	9	11	12	14
6.5	1.8718	1.8733	1.8749	1.8764	1.8779	1.8795	1.8810	1.8825	1.8840	1.8856	2	3	5	6	8	9	11	12	14
6.6	1.8871	1.8886	1.8901	1.8916	1.8931	1.8946	1.8961	1.8976	1.8991	1.9006	2	3	5	6	8	9	11	12	14
6.7	1.9021	1.9036	1.9051	1.9066	1.9081	1.9095	1.9110	1.9125	1.9140	1.9155	1	3	4	6	7	9	10	12	13
6.8	1.9169	1.9184	1.9199	1.9213	1.9228	1.9242	1.9257	1.9272	1.9286	1.9301	1	3	4	6	7	9	10	12	13
6.9	1.9315	1.9330	1.9344	1.9359	1.9373	1.9387	1.9402	1.9416	1.9430	1.9445	1	3	4	6	7	9	10	12	13
7.0	1.9459	1.9473	1.9488	1.9502	1.9516	1.9530	1.9544	1.9559	1.9573	1.9587	1	3	4	6	7	9	10	11	13
7.1	1.9601	1.9615	1.9629	1.9643	1.9657	1.9671	1.9685	1.9699	1.9713	1.9727	1	3	4	6	7	8	10	11	13
7.2	1.9741	1.9755	1.9769	1.9782	1.9796	1.9810	1.9824	1.9838	1.9851	1.9865	1	3	4	6	7	8	10	11	12
7.3	1.9879	1.9892	1.9906	1.9920	1.9933	1.9947	1.9961	1.9974	1.9988	2.0001	1	3	4	5	7	8	10	11	12
7.4	2.0015	2.0028	2.0042	2.0055	2.0069	2.0082	2.0096	2.0109	2.0122	2.0136	1	3	4	5	7	8	9	11	12
7.5	2.0149	2.0162	2.0176	2.0189	2.0202	2.0215	2.0229	2.0242	2.0255	2.0268	1	3	4	5	7	8	9	11	12
7.6	2.0281	2.0295	2.0308	2.0321	2.0334	2.0347	2.0360	2.0373	2.0386	2.0399	1	3	4	5	7	8	9	10	12
7.7	2.0412	2.0425	2.0438	2.0451	2.0464	2.0477	2.0490	2.0503	2.0516	2.0528	1	3	4	5	6	8	9	10	12
7.8	2.0541	2.0554	2.0567	2.0580	2.0592	2.0605	2.0618	2.0631	2.0643	2.0656	1	3	4	5	6	8	9	10	11
7.9	2.0669	2.0681	2.0694	2.0707	2.0719	2.0732	2.0744	2.0757	2.0769	2.0782	1	3	4	5	6	8	9	10	11
8.0	2.0794	2.0807	2.0819	2.0832	2.0844	2.0857	2.0869	2.0882	2.0894	2.0906	1	2	4	5	6	7	9	10	11
8.1	2.0919	2.0931	2.0943	2.0956	2.0968	2.0980	2.0992	2.1005	2.1017	2.1029	1	2	4	5	6	7	9	10	11
8.2	2.1041	2.1054	2.1066	2.1078	2.1090	2.1102	2.1114	2.1126	2.1138	2.1150	1	2	4	5	6	7	8	10	11
8.3	2.1163	2.1175	2.1187	2.1199	2.1211	2.1223	2.1235	2.1247	2.1258	2.1270	1	2	4	5	6	7	8	10	11
8.4	2.1282	2.1294	2.1306	2.1318	2.1330	2.1342	2.1353	2.1365	2.1377	2.1389	1	2	4	5	6	7	8	9	11
8.5	2.1401	2.1412	2.1424	2.1436	2.1448	2.1459	2.1471	2.1483	2.1494	2.1506	1	2	4	5	6	7	8	9	11
8.6	2.1518	2.1529	2.1541	2.1552	2.1564	2.1576	2.1587	2.1599	2.1610	2.1622	1	2	3	5	6	7	8	9	10
8.7	2.1633	2.1645	2.1656	2.1668	2.1679	2.1691	2.1702	2.1713	2.1725	2.1736	1	2	3	5	6	7	8	9	10
8.8	2.1748	2.1759	2.1770	2.1782	2.1793	2.1804	2.1815	2.1827	2.1838	2.1849	1	2	3	5	6	7	8	9	10
8.9	2.1861	2.1872	2.1883	2.1894	2.1905	2.1917	2.1928	2.1939	2.1950	2.1961	1	2	3	4	6	7	8	9	10
9.0	2.1972	2.1983	2.1994	2.2006	2.2017	2.2028	2.2039	2.2050	2.2061	2.2072	1	2	3	4	6	7	8	9	10
9.1	2.2083	2.2094	2.2105	2.2116	2.2127	2.2138	2.2148	2.2159	2.2170	2.2181	1	2	3	4	5	7	8	9	10
9.2	2.2192	2.2203	2.2214	2.2225	2.2235	2.2246	2.2257	2.2268	2.2279	2.2289	1	2	3	4	5	6	8	9	10
9.3	2.2300	2.2311	2.2322	2.2332	2.2343	2.2354	2.2364	2.2375	2.2386	2.2396	1	2	3	4	5	6	7	9	10
9.4	2.2407	2.2418	2.2428	2.2439	2.2450	2.2460	2.2471	2.2481	2.2492	2.2502	1	2	3	4	5	6	7	8	10
9.5	2.2513	2.2523	2.2534	2.2544	2.2555	2.2565	2.2576	2.2586	2.2597	2.2607	1	2	3	4	5	6	7	8	9
9.6	2.2618	2.2628	2.2638	2.2649	2.2659	2.2670	2.2680	2.2690	2.2701	2.2711	1	2	3	4	5	6	7	8	9
9.7	2.2721	2.2732	2.2742	2.2752	2.2762	2.2773	2.2783	2.2793	2.2803	2.2814	1	2	3	4	5	6	7	8	9
9.8	2.2824	2.2834	2.2844	2.2854	2.2865	2.2875	2.2885	2.2895	2.2905	2.2915	1	2	3	4	5	6	7	8	9
9.9	2.2925	2.2935	2.2946	2.2956	2.2966	2.2976	2.2986	2.2996	2.3006	2.3016	1	2	3	4	5	6	7	8	9

Natural Logarithms of 10^{-n}

n	1	2	3	4	5	6	7	8	9
$\ln 10^{-n}$	$\bar{3}.6974$	$\bar{5}.3948$	$\bar{7}.0922$	$\bar{10}.7897$	$\bar{12}.4871$	$\bar{14}.1845$	$\bar{17}.8819$	$\bar{19}.5793$	$\bar{21}.2767$

To find $\log_e 0.05361$

$$0.05361 = \frac{5.361}{100} = \frac{5.361}{10^2} = 5.361 \times 10^{-2}$$

$$\log_e 0.05361 = \log_e 5.361 + \log_e 10^{-2}$$
$$= 1.6792 + \bar{5}.3948 = \bar{3}.0740$$
$$= -3 + 0.0740 = -2.9260$$

81

Table of e^x

x	.00	.01	.02	.03	.04	.05	.06	.07	.08	.09
0.0	1.0000	1.0101	1.0202	1.0305	1.0408	1.0513	1.0618	1.0725	1.0833	1.0942
0.1	1.1052	1.1163	1.1275	1.1388	1.1503	1.1618	1.1735	1.1853	1.1972	1.2092
0.2	1.2214	1.2337	1.2461	1.2586	1.2712	1.2840	1.2969	1.3100	1.3231	1.3364
0.3	1.3499	1.3634	1.3771	1.3910	1.4049	1.4191	1.4333	1.4477	1.4623	1.4770
0.4	1.4918	1.5068	1.5220	1.5373	1.5527	1.5683	1.5841	1.6000	1.6161	1.6323
0.5	1.6487	1.6653	1.6820	1.6989	1.7160	1.7333	1.7507	1.7683	1.7860	1.8040
0.6	1.8221	1.8404	1.8589	1.8776	1.8965	1.9155	1.9348	1.9542	1.9739	1.9937
0.7	2.0138	2.0340	2.0544	2.0751	2.0959	2.1170	2.1383	2.1598	2.1815	2.2034
0.8	2.2255	2.2479	2.2705	2.2933	2.3164	2.3396	2.3632	2.3869	2.4109	2.4351
0.9	2.4596	2.4843	2.5093	2.5345	2.5600	2.5857	2.6117	2.6379	2.6645	2.6912
1.0	2.7183	2.7456	2.7732	2.8011	2.8292	2.8576	2.8864	2.9154	2.9447	2.9743
1.1	3.0042	3.0344	3.0649	3.0957	3.1268	3.1582	3.1899	3.2220	3.2544	3.2871
1.2	3.3201	3.3535	3.3872	3.4212	3.4556	3.4903	3.5254	3.5608	3.5966	3.6328
1.3	3.6693	3.7062	3.7434	3.7810	3.8190	3.8574	3.8962	3.9354	3.9749	4.0149
1.4	4.0552	4.0960	4.1371	4.1787	4.2207	4.2631	4.3060	4.3492	4.3929	4.4371
1.5	4.4817	4.5267	4.5722	4.6182	4.6646	4.7115	4.7588	4.8066	4.8550	4.9037
1.6	4.9530	5.0028	5.0531	5.1039	5.1552	5.2070	5.2593	5.3122	5.3656	5.4195
1.7	5.4739	5.5290	5.5845	5.6407	5.6973	5.7546	5.8124	5.8709	5.9299	5.9895
1.8	6.0496	6.1104	6.1719	6.2339	6.2965	6.3598	6.4237	6.4883	6.5535	6.6194
1.9	6.6859	6.7531	6.8210	6.8895	6.9588	7.0287	7.0993	7.1707	7.2427	7.3155
2.0	7.3891	7.4633	7.5383	7.6141	7.6906	7.7679	7.8460	7.9248	8.0045	8.0849
2.1	8.1662	8.2482	8.3311	8.4149	8.4994	8.5849	8.6711	8.7583	8.8463	8.9352
2.2	9.0250	9.1157	9.2073	9.2999	9.3933	9.4877	9.5831	9.6794	9.7767	9.8749
2.3	9.9742	10.074	10.176	10.278	10.381	10.486	10.591	10.697	10.805	10.913
2.4	11.023	11.134	11.246	11.359	11.473	11.588	11.705	11.822	11.941	12.061
2.5	12.183	12.305	12.429	12.554	12.680	12.807	12.936	13.066	13.197	13.330
2.6	13.464	13.599	13.736	13.874	14.013	14.154	14.296	14.440	14.585	14.732
2.7	14.880	15.029	15.180	15.333	15.487	15.643	15.800	15.959	16.119	16.281
2.8	16.445	16.610	16.777	16.945	17.116	17.288	17.462	17.637	17.814	17.993
2.9	18.174	18.357	18.541	18.728	18.916	19.106	19.298	19.492	19.688	19.886
3.0	20.086	20.287	20.491	20.697	20.905	21.115	21.327	21.542	21.758	21.977
3.1	22.198	22.421	22.646	22.874	23.104	23.336	23.571	23.808	24.047	24.288
3.2	24.533	24.779	25.028	25.280	25.534	25.790	26.050	26.311	26.576	26.843
3.3	27.113	27.385	27.660	27.938	28.219	28.503	28.789	29.079	29.371	29.666
3.4	29.964	30.265	30.569	30.877	31.187	31.500	31.817	32.137	32.460	32.786
3.5	33.115	33.448	33.784	34.124	34.467	34.813	35.163	35.517	35.874	36.234
3.6	36.598	36.966	37.338	37.713	38.092	38.475	38.861	39.252	39.646	40.045
3.7	40.447	40.854	41.264	41.679	42.098	42.521	42.948	43.380	43.816	44.256
3.8	44.701	45.150	45.604	46.063	46.525	46.993	47.465	47.942	48.424	48.911
3.9	49.402	49.899	50.400	50.907	51.419	51.935	52.457	52.985	53.517	54.055
4.0	54.598									

x	e^x	x	e^x	x	e^x	x	e^x	x	e^x	x	e^x
4.1	60.340	4.6	99.484	5.1	164.02	5.6	270.43	6.1	445.86	6.6	735.10
4.2	66.686	4.7	109.95	5.2	181.27	5.7	298.87	6.2	492.75	6.7	812.41
4.3	73.700	4.8	121.51	5.3	200.34	5.8	330.30	6.3	544.57	6.8	897.85
4.4	81.451	4.9	134.29	5.4	221.41	5.9	365.04	6.4	601.85	6.9	992.27
4.5	90.017	5.0	148.41	5.5	244.69	6.0	403.43	6.5	665.14	7.0	1096.63

Table of e^{-x}

x	.00	.01	.02	.03	.04	.05	.06	.07	.08	.09
0.0	1.0000	0.9900	0.9802	0.9704	0.9608	0.9512	0.9418	0.9324	0.9231	0.9139
0.1	0.9048	0.8958	0.8869	0.8781	0.8694	0.8607	0.8521	0.8437	0.8353	0.8270
0.2	0.8187	0.8106	0.8025	0.7945	0.7866	0.7788	0.7711	0.7634	0.7558	0.7483
0.3	0.7408	0.7334	0.7261	0.7189	0.7118	0.7047	0.6977	0.6907	0.6839	0.6771
0.4	0.6703	0.6637	0.6570	0.6505	0.6440	0.6376	0.6313	0.6250	0.6188	0.6126
0.5	0.6065	0.6005	0.5945	0.5886	0.5827	0.5769	0.5712	0.5655	0.5599	0.5543
0.6	0.5488	0.5434	0.5379	0.5326	0.5273	0.5220	0.5169	0.5117	0.5066	0.5016
0.7	0.4966	0.4916	0.4868	0.4819	0.4771	0.4724	0.4677	0.4630	0.4584	0.4538
0.8	0.4493	0.4449	0.4404	0.4360	0.4317	0.4274	0.4232	0.4190	0.4148	0.4107
0.9	0.4066	0.4025	0.3985	0.3946	0.3906	0.3867	0.3829	0.3791	0.3753	0.3716
1.0	0.3679	0.3642	0.3606	0.3570	0.3535	0.3499	0.3465	0.3430	0.3396	0.3362
1.1	0.3329	0.3296	0.3263	0.3230	0.3198	0.3166	0.3135	0.3104	0.3073	0.3042
1.2	0.3012	0.2982	0.2952	0.2923	0.2894	0.2865	0.2837	0.2808	0.2780	0.2753
1.3	0.2725	0.2698	0.2671	0.2645	0.2618	0.2592	0.2567	0.2541	0.2516	0.2491
1.4	0.2466	0.2441	0.2417	0.2393	0.2369	0.2346	0.2322	0.2299	0.2276	0.2254
1.5	0.2231	0.2209	0.2187	0.2165	0.2144	0.2122	0.2101	0.2080	0.2060	0.2039
1.6	0.2019	0.1999	0.1979	0.1959	0.1940	0.1920	0.1901	0.1882	0.1864	0.1845
1.7	0.1827	0.1809	0.1791	0.1773	0.1755	0.1738	0.1720	0.1703	0.1686	0.1670
1.8	0.1653	0.1637	0.1620	0.1604	0.1588	0.1572	0.1557	0.1541	0.1526	0.1511
1.9	0.1496	0.1481	0.1466	0.1451	0.1437	0.1423	0.1409	0.1395	0.1381	0.1367
2.0	0.1353	0.1340	0.1327	0.1313	0.1300	0.1287	0.1275	0.1262	0.1249	0.1237
2.1	0.1225	0.1212	0.1200	0.1188	0.1177	0.1165	0.1153	0.1142	0.1130	0.1119
2.2	0.1108	0.1097	0.1086	0.1075	0.1065	0.1054	0.1044	0.1033	0.1023	0.1013
2.3	0.1003	0.0993	0.0983	0.0973	0.0963	0.0954	0.0944	0.0935	0.0925	0.0916
2.4	0.0907	0.0898	0.0889	0.0880	0.0872	0.0863	0.0854	0.0846	0.0837	0.0829
2.5	0.0821	0.0813	0.0805	0.0797	0.0789	0.0781	0.0773	0.0765	0.0758	0.0750
2.6	0.0743	0.0735	0.0728	0.0721	0.0714	0.0707	0.0699	0.0693	0.0686	0.0679
2.7	0.0672	0.0665	0.0659	0.0652	0.0646	0.0639	0.0633	0.0627	0.0620	0.0614
2.8	0.0608	0.0602	0.0596	0.0590	0.0584	0.0578	0.0573	0.0567	0.0561	0.0556
2.9	0.0550	0.0545	0.0539	0.0534	0.0529	0.0523	0.0518	0.0513	0.0508	0.0503
3.0	0.0498	0.0493	0.0488	0.0483	0.0478	0.0474	0.0469	0.0464	0.0460	0.0455
3.1	0.0450	0.0446	0.0442	0.0437	0.0433	0.0429	0.0424	0.0420	0.0416	0.0412
3.2	0.0408	0.0404	0.0400	0.0396	0.0392	0.0388	0.0384	0.0380	0.0376	0.0373
3.3	0.0369	0.0365	0.0362	0.0358	0.0354	0.0351	0.0347	0.0344	0.0340	0.0337
3.4	0.0334	0.0330	0.0327	0.0324	0.0321	0.0317	0.0314	0.0311	0.0308	0.0305
3.5	0.0302	0.0299	0.0296	0.0293	0.0290	0.0287	0.0284	0.0282	0.0279	0.0276
3.6	0.0273	0.0271	0.0268	0.0265	0.0263	0.0260	0.0257	0.0255	0.0252	0.0250
3.7	0.0247	0.0245	0.0242	0.0240	0.0238	0.0235	0.0233	0.0231	0.0228	0.0226
3.8	0.0224	0.0221	0.0219	0.0217	0.0215	0.0213	0.0211	0.0209	0.0207	0.0204
3.9	0.0202	0.0200	0.0198	0.0196	0.0194	0.0193	0.0191	0.0189	0.0187	0.0185
4.0	0.0183									

e^{-x} for values of x greater than 4.0 may be found by using the table of natural logarithms and the reciprocal table.

To find $e^{-4.5361}$

If $y = e^{4.5361}$ then from the e^x table y is about 90.

Since 4.5361 is outside the range of values given in the natural logarithm tables, we use the table of natural logarithrims of 10^{+n}

We find that $\log_e 10 = 2.3026$

and hence $\log_e y = 2.3026 + 2.335$

$\qquad \log_e 9.333 = 2.2335$ (using the natural logarithm table)

Hence $e^{4.5361} = 9.333 \times 10 = 93.33$

$\qquad e^{-4.5361} = 0.01072$ (using the table of reciprocals)

Areas of Circles

Diam.	0.0	0.1	0.2	0.3	0.4	0.5	0.6	0.7	0.8	0.9
0	0.0	0.0079	0.0314	0.0707	0.1257	0.1963	0.2827	0.3848	0.5027	0.6362
1	0.7854	0.9503	1.1310	1.3273	1.5394	1.7671	2.0106	2.2698	2.5447	2.8353
2	3.1416	3.4636	3.8013	4.1548	4.5239	4.9087	5.3093	5.7256	6.1575	6.6052
3	7.0686	7.5477	8.0425	8.5530	9.0792	9.6211	10.179	10.752	11.341	11.946
4	12.566	13.203	13.854	14.522	15.205	15.904	16.619	17.349	18.096	18.857
5	19.635	20.428	21.237	22.062	22.902	23.758	24.630	25.518	26.421	27.340
6	28.274	29.225	30.191	31.172	32.170	33.183	34.212	35.257	36.317	37.393
7	38.485	39.592	40.715	41.854	43.008	44.179	45.365	46.566	47.784	49.017
8	50.266	51.530	52.810	54.106	55.418	56.745	58.088	59.447	60.821	62.211
9	63.617	65.039	66.476	67.929	69.398	70.882	72.382	73.898	75.430	76.977
10	78.540	80.118	81.713	83.323	84.949	86.590	88.247	89.920	91.609	93.313
11	95.033	96.769	98.520	100.29	102.07	103.87	105.68	107.51	109.36	111.22
12	113.10	114.99	116.90	118.82	120.76	122.72	124.69	126.68	128.68	130.70
13	132.73	134.78	136.85	138.93	141.03	143.14	145.27	147.41	149.57	151.75
14	153.94	156.15	158.37	160.61	162.86	165.13	167.42	169.72	172.03	174.37
15	176.72	179.08	181.46	183.85	186.27	188.69	191.13	193.59	196.07	198.56
16	201.06	203.58	206.12	208.67	211.24	213.82	216.42	219.04	221.67	224.32
17	226.98	229.66	232.35	235.06	237.79	240.53	243.28	246.06	248.85	251.65
18	254.47	257.30	260.16	263.02	265.90	268.80	271.72	274.65	277.59	280.55
19	283.53	286.52	289.53	292.55	295.59	298.65	301.72	304.80	307.91	311.03
20	314.16	317.31	320.47	323.65	326.85	330.06	333.29	336.54	339.79	343.07
21	346.36	349.67	352.99	356.33	359.68	363.05	366.44	369.84	373.25	376.68
22	380.13	383.60	387.08	390.57	394.08	397.61	401.15	404.71	408.28	411.87
23	415.48	419.10	422.73	426.38	430.05	433.74	437.44	441.15	444.88	448.63
24	452.39	456.17	459.96	463.77	467.59	471.44	475.29	479.16	483.05	486.95
25	490.88	494.81	498.76	502.73	506.71	510.71	514.72	518.75	522.79	526.85
26	530.93	535.02	539.13	543.25	547.39	551.55	555.72	559.90	564.10	568.32
27	572.56	576.80	581.07	585.35	589.65	593.96	598.28	602.63	606.99	611.36
28	615.75	620.16	624.58	629.02	633.47	637.94	642.42	646.92	651.44	655.97
29	660.52	665.08	669.66	674.26	678.87	684.49	688.13	692.79	697.47	702.15
30	706.86	711.58	716.31	721.07	725.83	730.62	735.42	740.23	745.06	749.91
31	754.77	759.65	764.54	769.45	774.37	779.31	784.27	789.24	794.23	799.23
32	804.25	809.28	814.33	819.40	824.48	829.58	834.69	839.82	844.96	850.12
33	855.30	860.49	865.70	870.92	876.16	881.41	886.68	891.97	897.27	902.59
34	907.92	913.27	918.63	924.01	929.41	934.82	940.25	945.69	951.15	956.62
35	962.12	967.62	973.14	978.68	984.23	989.80	995.38	1000.98	1006.6	1012.2
36	1017.9	1023.5	1029.2	1034.9	1040.6	1046.3	1052.1	1057.8	1063.6	1069.4
37	1075.2	1081.0	1086.9	1092.7	1098.6	1104.5	1110.4	1116.3	1122.2	1128.2
38	1134.1	1140.1	1146.1	1152.1	1158.1	1164.2	1170.2	1176.3	1182.4	1188.5
39	1194.6	1200.7	1206.9	1213.0	1219.2	1225.4	1231.6	1237.9	1244.1	1250.4
40	1256.6	1269.2	1269.2	1275.6	1281.9	1288.2	1294.6	1310.0	1307.4	1313.8
41	1320.3	1326.7	1333.2	1339.6	1346.1	1352.7	1359.2	1365.7	1372.3	1378.9
42	1385.4	1392.0	1398.7	1405.3	1412.0	1418.6	1425.3	1432.0	1438.7	1445.5
43	1452.2	1459.0	1465.7	1472.5	1479.3	1486.2	1493.0	1499.9	1506.7	1513.6
44	1520.5	1527.5	1534.4	1541.3	1548.3	1555.3	1562.3	1569.3	1576.3	1583.4
45	1590.4	1597.5	1604.6	1611.7	1618.8	1626.0	1633.1	1640.3	1647.5	1654.7
46	1661.9	1669.1	1676.4	1683.7	1690.9	1698.2	1705.5	1712.9	1720.2	1727.6
47	1734.9	1742.3	1749.7	1757.2	1764.6	1772.1	1779.5	1787.0	1794.5	1802.0
48	1809.6	1817.1	1824.7	1832.2	1839.8	1847.5	1855.1	1862.7	1870.4	1878.1
49	1885.7	1893.4	1901.2	1908.9	1916.7	1924.4	1932.2	1940.0	1947.8	1955.6
50	1963.5	1971.4	1979.2	1987.1	1995.0	2003.0	2010.9	2018.9	2026.8	2034.8

Area of circle of 18.6 mm diameter = 271.72 mm
Area of circle of 167 mm diameter.
Diameter = 16.7×10;   Area = $219.04 \times 100 = 21.904$ mm^2.

SIMPLE INTEREST

Appreciation of £1 for periods from 1 year to 24 years

Year	5%	6%	7%	8%	9%	10%	11%	12%	13%	14%
1	1·050	1·060	1·070	1·080	1·090	1·100	1·110	1·120	1·130	1·140
2	1·100	1·120	1·140	1·160	1·180	1·200	1·220	1·240	1·260	1·280
3	1·150	1·180	1·210	1·240	1·270	1·300	1·330	1·360	1·390	1·420
4	1·200	1·240	1·280	1·320	1·360	1·400	1·440	1·480	1·520	1·560
5	1·250	1·300	1·350	1·400	1·450	1·500	1·550	1·600	1·650	1·700
6	1·300	1·360	1·420	1·480	1·540	1·600	1·660	1·720	1·780	1·840
7	1·350	1·420	1·490	1·560	1·630	1·700	1·770	1·840	1·910	1·980
8	1·400	1·480	1·560	1·640	1·720	1·800	1·880	1·960	2·040	2·120
9	1·450	1·540	1·630	1·720	1·810	1·900	1·990	2·080	2·170	2·260
10	1·500	1·600	1·700	1·800	1·900	2·000	2·100	2·200	2·300	2·400
11	1·550	1·660	1·770	1·880	1·990	2·100	2·210	2·320	2·430	2·540
12	1·600	1·720	1·840	1·960	2·080	2·200	2·320	2·440	2·560	2·680
13	1·650	1·780	1·910	2·040	2·170	2·300	2·430	2·560	2·690	2·820
14	1·700	1·840	1·980	2·120	2·260	2·400	2·540	2·680	2·820	2·960
15	1·750	1·900	2·050	2·200	2·350	2·500	2·650	2·800	2·950	3·100
16	1·800	1·960	2·120	2·280	2·440	2·600	2·760	2·920	3·080	3·240
17	1·850	2·020	2·190	2·360	2·530	2·700	2·870	3·040	3·210	3·380
18	1·900	2·080	2·260	2·440	2·620	2·800	2·980	3·160	3·340	3·520
19	1·950	2·140	2·330	2·520	2·710	2·900	3·090	3·280	3·470	3·660
20	2·000	2·200	2·400	2·600	2·800	3·000	3·200	3·400	3·600	3·800
21	2·050	2·260	2·470	2·680	2·890	3·100	3·310	3·520	3·730	3·940
22	2·100	2·320	2·540	2·760	2·980	3·200	3·420	3·640	3·860	4·080
23	2·150	2·380	2·610	2·840	3·070	3·300	3·530	3·760	3·990	4·220
24	2·200	2·440	2·680	2·920	3·160	3·400	3·640	3·880	4·120	4·360

COMPOUND INTEREST

Appreciation of £1 for periods from 1 year to 24 years

Year	5%	6%	7%	8%	9%	10%	11%	12%	13%	14%
1	1·050	1·060	1·070	1·080	1·090	1·100	1·110	1·120	1·130	1·140
2	1·103	1·124	1·145	1·166	1·188	1·210	1·232	1·254	1·277	1·300
3	1·158	1·191	1·225	1·260	1·295	1·331	1·368	1·405	1·443	1·482
4	1·216	1·262	1·311	1·360	1·412	1·464	1·518	1·574	1·603	1·689
5	1·276	1·338	1·403	1·469	1·539	1·611	1·685	1·762	1·842	1·925
6	1·340	1·419	1·501	1·587	1·677	1·772	1·870	1·974	2·082	2·195
7	1·407	1·504	1·606	1·714	1·828	1·949	2·076	2·211	2·353	2·502
8	1·477	1·594	1·718	1·851	1·993	2·144	2·304	2·476	2·658	2·853
9	1·551	1·689	1·838	1·999	2·172	2·358	2·558	2·773	3·004	3·252
10	1·629	1·791	1·967	2·159	2·367	2·594	2·839	3·106	3·395	3·707
11	1·710	1·898	2·105	2·332	2·580	2·853	3·152	3·479	3·836	4·226
12	1·796	2·012	2·252	2·518	2·813	3·138	3·498	3·896	4·335	4·818
13	1·886	2·133	2·410	2·720	3·066	3·452	3·883	4·363	4·898	5·492
14	1·980	2·261	2·579	2·937	3·342	3·797	4·310	4·887	5·535	6·261
15	2·079	2·397	2·759	3·172	3·642	4·177	4·785	5·474	6·254	7·130
16	2·183	2·540	2·952	3·426	3·970	4·595	5·311	6·130	7·067	8·137
17	2·292	2·693	3·159	3·700	4·328	5·054	5·895	6·866	7·986	9·276
18	2·407	2·854	3·380	3·996	4·717	5·560	6·544	7·690	9·024	10·575
19	2·527	3·026	3·617	4·316	5·142	6·116	7·263	8·613	10·197	12·056
20	2·653	3·207	3·870	4·661	5·604	6·727	8·062	9·646	11·523	13·743
21	2·786	3·400	4·141	5·034	6·109	7·400	8·949	10·804	13·021	15·668
22	2·925	3·604	4·430	5·437	6·659	8·140	9·934	12·100	14·714	17·861
23	3·072	3·820	4·741	5·871	7·258	8·954	11·026	13·552	16·627	20·362
24	3·225	4·049	5·072	6·341	7·911	9·850	12·239	15·179	18·788	23·212

Normal Distribution

u	0.00	0.01	0.02	0.03	0.04	0.05	0.06	0.07	0.08	0.09
0.0	0.50000	0.49601	0.49202	0.48803	0.48405	0.48006	0.47608	0.47210	0.46812	0.46414
0.1	0.46017	0.45620	0.45224	0.44828	0.44433	0.44038	0.43644	0.43250	0.42858	0.42465
0.2	0.42074	0.41683	0.41294	0.40905	0.40517	0.40129	0.39743	0.39358	0.38974	0.38591
0.3	0.38209	0.37828	0.37448	0.37070	0.36693	0.36317	0.35942	0.35569	0.35197	0.34827
0.4	0.34458	0.34090	0.33724	0.33360	0.32997	0.32636	0.32276	0.31918	0.31561	0.31207
0.5	0.30854	0.30503	0.30153	0.29806	0.29460	0.29116	0.28774	0.28434	0.28096	0.27760
0.6	0.27425	0.27093	0.26763	0.26435	0.26109	0.25785	0.25463	0.25143	0.24825	0.24510
0.7	0.24196	0.23885	0.23576	0.23269	0.22965	0.22663	0.22363	0.22065	0.21770	0.21476
0.8	0.21186	0.20897	0.20611	0.20327	0.20045	0.19766	0.19489	0.19215	0.18943	0.18673
0.9	0.18406	0.18141	0.17879	0.17619	0.17361	0.17106	0.16853	0.16602	0.16354	0.16109
1.0	0.15866	0.15625	0.15386	0.15150	0.14917	0.14686	0.14457	0.14231	0.14007	0.13786
1.1	0.13567	0.13350	0.13136	0.12924	0.12714	0.12507	0.12302	0.12100	0.11900	0.11702
1.2	0.11507	0.11314	0.11123	0.10935	0.10749	0.10565	0.10383	0.10204	0.10027	0.09853
1.3	0.09680	0.09510	0.09342	0.09176	0.09012	0.08851	0.08692	0.08534	0.08379	0.08226
1.4	0.08076	0.07927	0.07780	0.07636	0.07493	0.07353	0.07215	0.07078	0.06944	0.06811
1.5	0.06681	0.06552	0.06426	0.06301	0.06178	0.06057	0.05938	0.05821	0.05705	0.05592
1.6	0.05480	0.05370	0.05262	0.05155	0.05050	0.04947	0.04846	0.04746	0.04648	0.04551
1.7	0.04457	0.04363	0.04272	0.04182	0.04093	0.04006	0.03920	0.03836	0.03754	0.03673
1.8	0.03593	0.03515	0.03438	0.03362	0.03288	0.03216	0.03144	0.03074	0.03005	0.02938
1.9	0.02872	0.02807	0.02743	0.02680	0.02619	0.02559	0.02500	0.02442	0.02385	0.02330
2.0	0.02275	0.02222	0.02169	0.02118	0.02068	0.02018	0.01970	0.01923	0.01876	0.01831
2.1	0.01786	0.01743	0.01700	0.01659	0.01618	0.01578	0.01539	0.01500	0.01463	0.01426
2.2	0.01390	0.01355	0.01321	0.01287	0.01255	0.01222	0.01191	0.01160	0.01130	0.01101
2.3	0.01072	0.01044	0.01017	0.00990	0.00964	0.00939	0.00914	0.00889	0.00866	0.00842
2.4	0.00820	0.00798	0.00776	0.00755	0.00734	0.00714	0.00695	0.00676	0.00657	0.00639
2.5	0.00621	0.00604	0.00587	0.00570	0.00554	0.00539	0.00523	0.00508	0.00494	0.00480
2.6	0.00466	0.00453	0.00440	0.00427	0.00415	0.00402	0.00391	0.00379	0.00368	0.00357
2.7	0.00347	0.00336	0.00326	0.00317	0.00307	0.00298	0.00289	0.00280	0.00272	0.00264
2.8	0.00256	0.00248	0.00240	0.00233	0.00226	0.00219	0.00212	0.00205	0.00199	0.00193
2.9	0.00187	0.00181	0.00175	0.00169	0.00164	0.00159	0.00154	0.00149	0.00144	0.00139
3.0	0.00135	0.00131	0.00126	0.00122	0.00118	0.00114	0.00111	0.00107	0.00104	0.00100
3.1	0.00097	0.00094	0.00090	0.00087	0.00084	0.00082	0.00079	0.00076	0.00074	0.00071
3.2	0.00069	0.00066	0.00064	0.00062	0.00060	0.00058	0.00056	0.00054	0.00052	0.00050
3.3	0.00048	0.00047	0.00045	0.00043	0.00042	0.00040	0.00039	0.00038	0.00036	0.00035
3.4	0.00034	0.00032	0.00031	0.00030	0.00029	0.00028	0.00027	0.00026	0.00025	0.00024
3.5	0.00023	0.00022	0.00022	0.00021	0.00020	0.00019	0.00019	0.00018	0.00017	0.00017
3.6	0.00016	0.00015	0.00015	0.00014	0.00014	0.00013	0.00013	0.00012	0.00012	0.00011
3.7	0.00011	0.00010	0.00010	0.00010	0.00009	0.00009	0.00008	0.00008	0.00008	0.00008
3.8	0.00007	0.00007	0.00007	0.00006	0.00006	0.00006	0.00006	0.00005	0.00005	0.00005
3.9	0.00005	0.00005	0.00004	0.00004	0.00004	0.00004	0.00004	0.00004	0.00003	0.00003

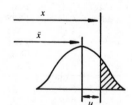

$$u = \frac{x - \bar{x}}{\sigma}$$

x = value of the variate
$\bar{x}$ = arithmetic mean
σ = standard deviation

Binomial Coefficients

r:	0	1	2	3	4	5	6	7	8	9	10
n=											
1	1	1									
2	1	2	1								
3	1	3	3	1							
4	1	4	6	4	1						
5	1	5	10	10	5	1					
6	1	6	15	20	15	6	1				
7	1	7	21	35	35	21	7	1			
8	1	8	28	56	70	56	28	8	1		
9	1	9	36	84	126	126	84	36	9	1	
10	1	10	45	120	210	252	210	120	45	10	1
11	1	11	55	165	330	462	462	330	165	55	11
12	1	12	66	220	495	792	924	792	495	220	66
13	1	13	78	286	715	1287	1716	1716	1287	715	286
14	1	14	91	364	1001	2002	3003	3432	3003	2002	1001
15	1	15	105	455	1365	3003	5005	6435	6435	5005	3003
16	1	16	120	560	1820	4368	8008	11440	12870	11440	8008
17	1	17	136	680	2380	6188	12376	19448	24310	24310	19448
18	1	18	153	816	3060	8568	18564	31824	43758	48620	43758
19	1	19	171	969	3876	11628	27132	50388	75582	92378	92378
20	1	20	190	1140	4845	15504	38760	77520	125970	167960	184756

$(a+b)^5 = a^5 + 5a^4b + 10a^3b^2 + 10a^2b^3 + 5ab^4 + b^5.$

Note that the coefficients are symmetrical. Thus

$$(a+b)^{15} = a^{15} + 15a^{14}b + 105a^{13}b^2 + 455a^{12}b^3 + 1365a^{11}b^4 + 3003a^{10}b^5$$
$$+ 5005a^9b^6 + 6435a^8b^7 + 6435a^7b^8 + 5005a^6b^9 + 3003a^5b^{10}$$
$$+ 1365a^4b^{11} + 455a^3b^{12} + 105a^2b^{13} + 15ab^{14} + b^{15}.$$

The Poisson Distribution

x	0.1	0.2	0.3	0.4	0.5	0.6	0.7	0.8	0.9	1.0
0	0.9048	0.8187	0.7408	0.6703	0.6065	0.5488	0.4966	0.4493	0.4066	0.3679
1	0.0905	0.1637	0.2222	0.2681	0.3033	0.3293	0.3476	0.3595	0.3659	0.3679
2	0.0045	0.0164	0.0333	0.0536	0.0758	0.0988	0.1217	0.1438	0.1647	0.1839
3	0.0002	0.0011	0.0033	0.0072	0.0126	0.0198	0.0284	0.0383	0.0494	0.0613
4	0.0000	0.0001	0.0002	0.0007	0.0016	0.0030	0.0050	0.0077	0.0111	0.0153
5	0.0000	0.0000	0.0000	0.0001	0.0002	0.0004	0.0007	0.0012	0.0020	0.0031
6	0.0000	0.0000	0.0000	0.0000	0.0000	0.0000	0.0001	0.0002	0.0003	0.0005
7	0.0000	0.0000	0.0000	0.0000	0.0000	0.0000	0.0000	0.0000	0.0000	0.0001

x	1.1	1.2	1.3	1.4	1.5	1.6	1.7	1.8	1.9	2.0
0	0.3329	0.3012	0.2725	0.2466	0.2231	0.2019	0.1827	0.1653	0.1496	0.1353
1	0.3662	0.3614	0.3543	0.3452	0.3347	0.3230	0.3106	0.2975	0.2842	0.2707
2	0.2014	0.2169	0.2303	0.2417	0.2510	0.2584	0.2640	0.2678	0.2700	0.2707
3	0.0738	0.0867	0.0998	0.1128	0.1255	0.1378	0.1496	0.1607	0.1710	0.1804
4	0.0203	0.0260	0.0324	0.0395	0.0471	0.0551	0.0636	0.0723	0.0812	0.0902
5	0.0045	0.0062	0.0084	0.0111	0.0141	0.0176	0.0216	0.0260	0.0309	0.0361
6	0.0008	0.0012	0.0018	0.0026	0.0035	0.0047	0.0061	0.0078	0.0098	0.0120
7	0.0001	0.0002	0.0003	0.0005	0.0008	0.0011	0.0015	0.0020	0.0027	0.0034
8	0.0000	0.0000	0.0001	0.0001	0.0001	0.0002	0.0003	0.0005	0.0006	0.0009
9	0.0000	0.0000	0.0000	0.0000	0.0000	0.0000	0.0001	0.0001	0.0001	0.0002

x	2.1	2.2	2.3	2.4	2.5	2.6	2.7	2.8	2.9	3.0
0	0.1225	0.1108	0.1003	0.0907	0.0821	0.0743	0.0672	0.0608	0.0550	0.0498
1	0.2572	0.2438	0.2306	0.2177	0.2052	0.1931	0.1815	0.1703	0.1596	0.1494
2	0.2700	0.2681	0.2652	0.2613	0.2565	0.2510	0.2450	0.2384	0.2314	0.2240
3	0.1890	0.1966	0.2033	0.2090	0.2138	0.2176	0.2205	0.2225	0.2237	0.2240
4	0.0992	0.1082	0.1169	0.1254	0.1336	0.1414	0.1488	0.1557	0.1622	0.1680
5	0.0417	0.0476	0.0538	0.0602	0.0668	0.0735	0.0804	0.0872	0.0940	0.1008
6	0.0146	0.0174	0.0206	0.0241	0.0278	0.0319	0.0362	0.0407	0.0455	0.0504
7	0.0044	0.0055	0.0068	0.0083	0.0099	0.0118	0.0139	0.0163	0.0188	0.0216
8	0.0011	0.0015	0.0019	0.0025	0.0031	0.0038	0.0047	0.0057	0.0068	0.0081
9	0.0003	0.0004	0.0005	0.0007	0.0009	0.0011	0.0014	0.0018	0.0022	0.0027
10	0.0001	0.0001	0.0001	0.0002	0.0002	0.0003	0.0004	0.0005	0.0006	0.0008
11	0.0000	0.0000	0.0000	0.0000	0.0000	0.0001	0.0001	0.0001	0.0002	0.0002
12	0.0000	0.0000	0.0000	0.0000	0.0000	0.0000	0.0000	0.0000	0.0000	0.0001

Entries in the table give the probabilities that an event will occur x times when the average number of occurrences is λ.

The Poisson Distribution

x	3.1	3.2	3.3	3.4	3.5	3.6	3.7	3.8	3.9	4.0
0	0.0450	0.0408	0.0369	0.0334	0.0302	0.0273	0.0247	0.0224	0.0202	0.0183
1	0.1397	0.1304	0.1217	0.1135	0.1057	0.0984	0.0915	0.0850	0.0789	0.0733
2	0.2165	0.2087	0.2008	0.1929	0.1850	0.1771	0.1692	0.1615	0.1539	0.1465
3	0.2237	0.2226	0.2209	0.2186	0.2158	0.2125	0.2087	0.2046	0.2001	0.1954
4	0.1734	0.1781	0.1823	0.1858	0.1888	0.1912	0.1931	0.1944	0.1951	0.1954
5	0.1075	0.1140	0.1203	0.1264	0.1322	0.1377	0.1429	0.1477	0.1522	0.1563
6	0.0555	0.0608	0.0662	0.0716	0.0771	0.0826	0.0881	0.0936	0.0989	0.1042
7	0.0246	0.0278	0.0312	0.0348	0.0385	0.0425	0.0466	0.0508	0.0551	0.0595
8	0.0095	0.0111	0.0129	0.0148	0.0169	0.0191	0.0215	0.0241	0.0269	0.0298
9	0.0033	0.0040	0.0047	0.0056	0.0066	0.0076	0.0089	0.0102	0.0116	0.0132
10	0.0010	0.0013	0.0016	0.0019	0.0023	0.0028	0.0033	0.0039	0.0045	0.0053
11	0.0003	0.0004	0.0005	0.0006	0.0007	0.0009	0.0011	0.0013	0.0016	0.0019
12	0.0001	0.0001	0.0001	0.0002	0.0002	0.0003	0.0003	0.0004	0.0005	0.0006
13	0.0000	0.0000	0.0000	0.0000	0.0001	0.0001	0.0001	0.0001	0.0002	0.0002
14	0.0000	0.0000	0.0000	0.0000	0.0000	0.0000	0.0000	0.0000	0.0000	0.0001

x	4.1	4.2	4.3	4.4	4.5	4.6	4.7	4.8	4.9	5.0
0	0.0166	0.0150	0.0136	0.0123	0.0111	0.0101	0.0091	0.0082	0.0074	0.0067
1	0.0679	0.0630	0.0583	0.0540	0.0500	0.0462	0.0427	0.0395	0.0365	0.0337
2	0.1393	0.1323	0.1254	0.1188	0.1125	0.1063	0.1005	0.0948	0.0894	0.0842
3	0.1904	0.1852	0.1798	0.1743	0.1687	0.1631	0.1574	0.1517	0.1460	0.1404
4	0.1951	0.1944	0.1933	0.1917	0.1898	0.1875	0.1849	0.1820	0.1789	0.1755
5	0.1600	0.1633	0.1662	0.1687	0.1708	0.1725	0.1738	0.1747	0.1753	0.1755
6	0.1093	0.1143	0.1191	0.1237	0.1281	0.1323	0.1362	0.1398	0.1432	0.1462
7	0.0640	0.0686	0.0732	0.0778	0.0824	0.0869	0.0914	0.0959	0.1002	0.1044
8	0.0328	0.0360	0.0393	0.0428	0.0463	0.0500	0.0537	0.0575	0.0614	0.0653
9	0.0150	0.0168	0.0188	0.0209	0.0232	0.0255	0.0280	0.0307	0.0334	0.0363
10	0.0061	0.0071	0.0081	0.0092	0.0104	0.0118	0.0132	0.0147	0.0164	0.0181
11	0.0023	0.0027	0.0032	0.0037	0.0043	0.0049	0.0056	0.0064	0.0073	0.0082
12	0.0008	0.0009	0.0011	0.0014	0.0016	0.0019	0.0022	0.0026	0.0030	0.0034
13	0.0002	0.0003	0.0004	0.0005	0.0006	0.0007	0.0008	0.0009	0.0011	0.0013
14	0.0001	0.0001	0.0001	0.0001	0.0002	0.0002	0.0003	0.0003	0.0004	0.0005
15	0.0000	0.0000	0.0000	0.0000	0.0001	0.0001	0.0001	0.0001	0.0001	0.0002

Entries in the table give the probabilities that an event will occur x times when the average number of occurrences is λ.

Percentage Points of the *t* Distribution

One sided test

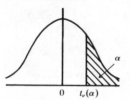

$$Pr(T_\nu > t_\nu(\alpha)) = \alpha,$$
for ν degrees of freedom

Two sided test

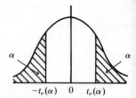

$$Pr(T_\nu > t_\nu(\alpha) \text{ and } T_\nu < -t_\nu(\alpha)) = 2\alpha$$
for ν degrees of freedom.

ν	$\alpha=0.4$ $2\alpha=0.8$	0.25 0.5	0.1 0.2	0.05 0.1	0.025 0.05	0.01 0.02	0.005 0.01	0.0025 0.005	0.001 0.002	0.0005 0.001
1	0.325	1.000	3.078	6.314	12.706	31.821	63.657	127.320	318.310	636.620
2	0.289	0.816	1.886	2.920	4.303	6.965	9.925	14.089	22.327	31.598
3	0.277	0.765	1.638	2.353	3.182	4.541	5.841	7.453	10.214	12.924
4	0.271	0.741	1.533	2.132	2.776	3.747	4.604	5.598	7.173	8.610
5	0.267	0.727	1.476	2.015	2.571	3.365	4.032	4.773	5.893	6.869
6	0.265	0.718	1.440	1.943	2.447	3.143	3.707	4.317	5.208	5.959
7	0.263	0.711	1.415	1.895	2.365	2.998	3.499	4.029	4.785	5.408
8	0.262	0.706	1.397	1.860	2.306	2.896	3.355	3.833	4.501	5.041
9	0.261	0.703	1.383	1.833	2.262	2.821	3.250	3.690	4.297	4.781
10	0.260	0.700	1.372	1.812	2.228	2.764	3.169	3.581	4.144	4.587
11	0.260	0.697	1.363	1.796	2.201	2.718	3.106	3.497	4.025	4.437
12	0.259	0.695	1.356	1.782	2.179	2.681	3.055	3.428	3.930	4.318
13	0.259	0.694	1.350	1.771	2.160	2.650	3.012	3.372	3.852	4.221
14	0.258	0.692	1.345	1.761	2.145	2.624	2.977	3.326	3.787	4.140
15	0.258	0.691	1.341	1.753	2.131	2.602	2.947	3.286	3.733	4.073
16	0.258	0.690	1.337	1.746	2.120	2.583	2.921	3.252	3.686	4.015
17	0.257	0.689	1.333	1.740	2.110	2.567	2.898	3.222	3.646	3.965
18	0.257	0.688	1.330	1.734	2.101	2.552	2.878	3.197	3.610	3.922
19	0.257	0.688	1.328	1.729	2.093	2.539	2.861	3.174	3.579	3.883
20	0.257	0.687	1.325	1.725	2.086	2.528	2.845	3.153	3.552	3.850
21	0.257	0.686	1.323	1.721	2.080	2.518	2.831	3.135	3.527	3.819
22	0.256	0.686	1.321	1.717	2.074	2.508	2.819	3.119	3.505	3.792
23	0.256	0.685	1.319	1.714	2.069	2.500	2.807	3.104	3.485	3.767
24	0.256	0.685	1.318	1.711	2.064	2.492	2.797	3.091	3.467	3.745
25	0.256	0.684	1.316	1.708	2.060	2.485	2.787	3.078	3.450	3.725
26	0.256	0.684	1.315	1.706	2.056	2.479	2.779	3.067	3.435	3.707
27	0.256	0.684	1.314	1.703	2.052	2.473	2.771	3.057	3.421	3.690
28	0.256	0.683	1.313	1.701	2.048	2.467	2.763	3.047	3.408	3.674
29	0.256	0.683	1.311	1.699	2.045	2.462	2.756	3.038	3.396	3.659
30	0.256	0.683	1.310	1.697	2.042	2.457	2.750	3.030	3.385	3.646
40	0.255	0.681	1.303	1.684	2.021	2.423	2.704	2.971	3.307	3.551
60	0.254	0.679	1.296	1.671	2.000	2.390	2.660	2.915	3.232	3.460
120	0.254	0.677	1.289	1.658	1.980	2.358	2.617	2.860	3.160	3.373
∞	0.253	0.674	1.282	1.645	1.960	2.326	2.576	2.807	3.090	3.291

Control Chart Factors for the Sample Average

If the standard deviation, σ, is the measure of dispersion, the limits are obtained as

$$\text{average} \pm (\hat{\sigma} \times \text{appropriate tabulated factor}),$$

where $\hat{\sigma}$ is an estimate of σ.

Sample Size	One-sided percentage level					
	0.1	0.5	1.0	2.5	5.0	10.0
2	2.19	1.82	1.64	1.39	1.16	0.91
3	1.78	1.49	1.34	1.13	0.95	0.74
4	1.55	1.29	1.16	0.98	0.82	0.64
5	1.38	1.15	1.04	0.88	0.74	0.57
6	1.26	1.05	0.95	0.80	0.67	0.52
7	1.17	0.97	0.88	0.74	0.62	0.48
8	1.09	0.91	0.82	0.69	0.58	0.45
9	1.03	0.86	0.78	0.65	0.55	0.43
10	0.98	0.81	0.74	0.62	0.52	0.41
11	0.93	0.78	0.70	0.59	0.50	0.39
12	0.89	0.74	0.67	0.57	0.47	0.37
13	0.86	0.71	0.65	0.54	0.46	0.36
14	0.83	0.69	0.62	0.52	0.44	0.34
15	0.80	0.67	0.60	0.51	0.42	0.33
16	0.77	0.64	0.58	0.49	0.41	0.32
17	0.75	0.62	0.56	0.48	0.40	0.31
18	0.73	0.61	0.55	0.46	0.39	0.30
19	0.71	0.59	0.53	0.45	0.38	0.29
20	0.69	0.58	0.52	0.44	0.37	0.29

When the measure of dispersion is range, ω, the limits are obtained as

$$\text{average} \pm (\hat{\omega} \times \text{appropriate tabulated factor}),$$

where $\hat{\omega}$ is an estimate of ω.

Sample Size	One-sided percentage level					
	0.1	0.5	1.0	2.5	5.0	10.0
2	1.94	1.61	1.46	1.23	1.03	0.80
3	1.05	0.88	0.79	0.67	0.56	0.44
4	0.75	0.63	0.56	0.48	0.40	0.31
5	0.59	0.50	0.45	0.38	0.32	0.25
6	0.50	0.41	0.37	0.32	0.26	0.21
7	0.43	0.36	0.33	0.27	0.23	0.18
8	0.38	0.32	0.29	0.24	0.20	0.16
9	0.35	0.29	0.26	0.22	0.18	0.14
10	0.32	0.26	0.24	0.20	0.17	0.13
11	0.29	0.24	0.22	0.19	0.16	0.12
12	0.27	0.23	0.21	0.17	0.15	0.11
13	0.26	0.21	0.19	0.16	0.14	0.11
14	0.24	0.20	0.18	0.15	0.13	0.10
15	0.23	0.19	0.17	0.15	0.12	0.10
16	0.22	0.18	0.16	0.14	0.12	0.09
17	0.21	0.17	0.16	0.13	0.11	0.09
18	0.20	0.17	0.15	0.13	0.11	0.08
19	0.19	0.16	0.14	0.12	0.10	0.08
20	0.18	0.15	0.14	0.12	0.10	0.08

Percentage Points of the X² Distribution

The values tabulated are $\chi_\nu^2(\alpha)$, where
$Pr(\chi_\nu^2 > \chi_\nu^2(\alpha)) = \alpha$, for ν degrees of freedom.

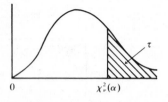

$0 \qquad\qquad\qquad \chi_\nu^2(\alpha)$

α / ν	0.250	0.100	0.050	0.025	0.010	0.005	0.001
1	1.32330	2.70554	3.84146	5.02389	6.63490	7.87944	10.828
2	2.77259	4.60517	5.99146	7.37776	9.21034	10.5966	13.816
3	4.10834	6.25139	7.81473	9.34840	11.3449	12.8382	16.266
4	5.38527	7.77944	9.48773	11.1433	13.2767	14.8603	18.467
5	6.62568	9.23636	11.0705	12.8325	15.0863	16.7496	20.515
6	7.84080	10.6446	12.5916	14.4494	16.8119	18.5476	22.458
7	9.03715	12.0170	14.0671	16.0128	18.4753	20.2777	24.322
8	10.2189	13.3616	15.5073	17.5345	20.0902	21.9550	26.125
9	11.3888	14.6837	16.9190	19.0228	21.6660	23.5894	27.877
10	12.5489	15.9872	18.3070	20.4832	23.2093	25.1882	29.588
11	13.7007	17.2750	19.6751	21.9200	24.7250	26.7568	31.264
12	14.8454	18.5493	21.0261	23.3367	26.2170	28.2995	32.909
13	15.9839	19.8119	22.3620	24.7356	27.6882	29.8195	34.528
14	17.1169	21.0641	23.6848	26.1189	29.1412	31.3194	36.123
15	18.2451	22.3071	24.9958	27.4884	30.5779	32.8013	37.697
16	19.3689	23.5418	26.2962	28.8454	31.9999	34.2672	39.252
17	20.4887	24.7690	27.5871	30.1910	33.4087	35.7185	40.790
18	21.6049	25.9894	28.8693	31.5264	34.8053	37.1565	42.312
19	22.7178	27.2036	30.1435	32.8523	36.1909	38.5823	43.820
20	23.8277	28.4120	31.4104	34.1696	37.5662	39.9968	45.315
21	24.9348	29.6151	32.6706	35.4789	38.9322	41.4011	46.797
22	26.0393	30.8133	33.9244	36.7807	40.2894	42.7957	48.268
23	27.1413	32.0069	35.1725	38.0756	41.6384	44.1813	49.728
24	28.2412	33.1962	36.4150	39.3641	42.9798	45.5585	51.179
25	29.3389	34.3816	37.6525	40.6465	44.3141	46.9279	52.618
26	30.4346	35.5632	38.8851	41.9232	45.6417	48.2899	54.052
27	31.5284	36.7412	40.1133	43.1945	46.9629	49.6449	55.476
28	32.6205	37.9159	41.3371	44.4608	48.2782	50.9934	56.892
29	33.7109	39.0875	42.5570	45.7223	49.5879	52.3356	58.301
30	34.7997	40.2560	43.7730	46.9792	50.8922	53.6720	59.703
40	45.6160	51.8051	55.7585	59.3417	63.6907	66.7660	73.402
50	56.3336	63.1671	67.5048	71.4202	76.1539	79.4900	86.661
60	66.9815	74.3970	79.0819	83.2977	88.3794	91.9517	99.607
70	77.5767	85.5270	90.5312	95.0232	100.425	104.215	112.317
80	88.1303	96.5782	101.879	106.629	112.329	116.321	124.839
90	98.6499	107.565	113.145	118.136	124.116	128.299	137.208
100	109.141	118.498	124.342	129.561	135.807	140.169	149.449

Percentage Points of the X² Distribution

For $\nu > 30$ take $\chi^2_\nu(\alpha) = \nu\left[1 - \dfrac{2}{9\nu} + u_\alpha\sqrt{\dfrac{2}{9\nu}}\right]^3$ where u_α is such that $Pr(U > u_\alpha) = \alpha$, and $U \sim N(0, 1)$.

0.995	0.990	0.975	0.950	0.900	0.750	0.500	α / ν
392704.10⁻¹⁰	157088.10⁻⁹	982069.10⁻⁹	393214.10⁻⁸	0.0157908	0.1015308	0.454936	1
0.0100251	0.0201007	0.0506356	0.102587	0.210721	0.575364	1.38629	2
0.0717218	0.114832	0.215795	0.351846	0.584374	1.212534	2.36597	3
0.206989	0.297109	0.484419	0.710723	1.063623	1.92256	3.35669	4
0.411742	0.554298	0.831212	1.145476	1.61031	2.67460	4.35146	5
0.675727	0.872090	1.23734	1.63538	2.20413	3.45460	5.34812	6
0.989256	1.239043	1.68987	2.16735	2.83311	4.25485	6.34581	7
1.34441	1.64650	2.17973	2.73264	3.48954	5.07064	7.34412	8
1.73493	2.08790	2.70039	3.32511	4.16816	5.89883	8.34283	9
2.15586	2.55821	3.24697	3.94030	4.86518	6.73720	9.34182	10
2.60322	3.05348	3.81575	4.57481	5.57778	7.58414	10.3410	11
3.07382	3.57057	4.40379	5.22603	6.30380	8.43842	11.3403	12
3.56503	4.10692	5.00875	5.89186	7.04150	9.29907	12.3398	13
4.07467	4.66043	5.62873	6.57063	7.78953	10.1653	13.3393	14
4.60092	5.22935	6.26214	7.26094	8.54676	11.0365	14.3389	15
5.14221	5.81221	6.90766	7.96165	9.31224	11.9122	15.3385	16
5.69722	6.40776	7.56419	8.67176	10.0852	12.7919	16.3382	17
6.26480	7.01491	8.23075	9.39046	10.8649	13.6753	17.3379	18
6.84397	7.63273	8.90652	10.1170	11.6509	14.5620	18.3377	19
7.43384	8.26040	9.59078	10.8508	12.4426	15.4518	19.3374	20
8.03365	8.89720	10.28293	11.5913	13.2396	16.3444	20.3372	21
8.64272	9.54249	10.9823	12.3380	14.0415	17.2396	21.3370	22
9.26043	10.19567	11.6886	13.0905	14.8480	18.1373	22.3369	23
9.88623	10.8564	12.4012	13.8484	15.6587	19.0373	23.3367	24
10.5197	11.5240	13.1197	14.6114	16.4734	19.9393	24.3366	25
11.1602	12.1981	13.8439	15.3792	17.2919	20.8434	25.3365	26
11.8076	12.8785	14.5734	16.1514	18.1139	21.7494	26.3363	27
12.4613	13.5647	15.3079	16.9279	18.9392	22.6572	27.3362	28
13.1211	14.2565	16.0471	17.7084	19.7677	23.5666	28.3361	29
13.7867	14.9535	16.7908	18.4927	20.5992	24.4776	29.3360	30
20.7065	22.1643	24.4330	26.5093	29.0505	33.6603	39.3353	40
27.9907	29.7067	32.3574	34.7643	37.6886	42.9421	49.3349	50
35.5345	37.4849	40.4817	43.1880	46.4589	52.2938	59.3347	60
43.2752	45.4417	48.7576	51.7393	55.3289	61.6983	69.3345	70
51.1719	53.5401	57.1532	60.3915	64.2778	71.1445	79.3343	80
59.1963	61.7541	65.6466	69.1260	73.2911	80.6247	89.3342	90
67.3276	70.0649	74.2219	77.9295	82.3581	90.1332	99.3341	100

Control Chart Factors for the Sample Range Based on the Standard Deviation

The limits are obtained by multiplying the estimate of the standard deviation* by the appropriate tabulated factors.

Sample Size	Lower percentage factors						Upper percentage factors					
	0.1	0.5	1.0	2.5	5.0	10.0	10.0	5.0	2.5	1.0	0.5	0.1
2	0.00	0.01	0.02	0.04	0.09	0.18	2.33	2.77	3.17	3.64	3.97	4.65
3	0.06	0.13	0.19	0.30	0.43	0.62	2.90	3.31	3.68	4.12	4.42	5.06
4	0.20	0.34	0.43	0.59	0.76	0.98	3.24	3.63	3.98	4.40	4.69	5.31
5	0.37	0.55	0.67	0.85	1.03	1.26	3.48	3.86	4.20	4.60	4.89	5.48
6	0.53	0.75	0.87	1.07	1.25	1.49	3.66	4.03	4.36	4.76	5.03	5.62
7	0.69	0.92	1.05	1.25	1.44	1.68	3.81	4.17	4.49	4.88	5.15	5.73
8	0.83	1.08	1.20	1.41	1.60	1.84	3.93	4.29	4.60	4.99	5.25	5.82
9	0.97	1.21	1.34	1.55	1.74	1.97	4.04	4.39	4.70	5.08	5.34	5.90
10	1.08	1.33	1.47	1.67	1.86	2.09	4.13	4.47	4.78	5.16	5.42	5.97
11	1.19	1.45	1.58	1.78	1.97	2.20	4.21	4.55	4.86	5.23	5.49	6.04
12	1.29	1.55	1.68	1.88	2.07	2.30	4.28	4.62	4.92	5.29	5.55	6.09
13	1.39	1.64	1.77	1.98	2.16	2.39	4.35	4.68	4.99	5.35	5.60	6.14
14	1.47	1.72	1.86	2.06	2.24	2.47	4.41	4.74	5.04	5.40	5.65	6.19
15	1.55	1.80	1.93	2.14	2.32	2.54	4.47	4.80	5.09	5.45	5.70	6.23
16	1.62	1.88	2.01	2.21	2.39	2.61	4.52	4.85	5.14	5.49	5.74	6.27
17	1.69	1.94	2.07	2.27	2.45	2.67	4.57	4.89	5.18	5.54	5.78	6.31
18	1.76	2.01	2.14	2.34	2.52	2.73	4.61	4.93	5.22	5.57	5.82	6.35
19	1.82	2.07	2.20	2.39	2.57	2.79	4.65	4.97	5.26	5.61	5.86	6.38
20	1.88	2.12	2.25	2.45	2.63	2.84	4.69	5.01	5.30	5.65	5.89	6.41

*If the measure of dispersion is range, then first convert this to an estimate of the standard deviation using the table below.

Conversion of Range to Standard Deviation

Multiply the range by the appropriate tabulated factor.

Sample Size	Conversion Factor	Sample Size	Conversion Factor	Sample Size	Conversion Factor	Sample Size	Conversion Factor	Sample Size	Conversion Factor
2	0.8862	6	0.3946	10	0.3249	14	0.2935	18	0.2747
3	0.5908	7	0.3698	11	0.3152	15	0.2880	19	0.2711
4	0.4857	8	0.3512	12	0.3069	16	0.2831	20	0.2677
5	0.4299	9	0.3367	13	0.2998	17	0.2787		

Critical Values of the Spearman Rank Correlation Coefficient

n	Significance level (one-tailed test)	
	0.05	0.01
4	1.000	
5	0.900	1.000
6	0.829	0.943
7	0.714	0.893
8	0.643	0.833
9	0.600	0.783
10	0.564	0.746
12	0.506	0.712
14	0.456	0.645
16	0.425	0.601
18	0.399	0.564
20	0.377	0.534
22	0.359	0.508
24	0.343	0.485
26	0.329	0.465
28	0.317	0.448
30	0.306	0.432

The Distribution of the Kendall Rank Correlation Coefficient

S_t	Values of n				S_t	Values of n		
	4	5	8	9		6	7	10
0	0.625	0.592	0.548	0.540	1	0.500	0.500	0.500
2	0.375	0.408	0.452	0.460	3	0.360	0.386	0.431
4	0.167	0.242	0.360	0.381	5	0.235	0.281	0.364
6	0.042	0.117	0.274	0.306	7	0.136	0.191	0.300
8		0.042	0.199	0.238	9	0.068	0.119	0.242
10		0.0083	0.138	0.179	11	0.028	0.068	0.190
12			0.089	0.130	13	0.0083	0.035	0.146
14			0.054	0.090	15	0.0014	0.015	0.108
16			0.031	0.060	17		0.0054	0.078
18			0.016	0.038	19		0.0014	0.054
20			0.0071	0.022	21		0.0002	0.036
22			0.0028	0.012	23			0.023
24			0.0009	0.0063	25			0.014
26			0.0002	0.0029	27			0.0083
28				0.0012	29			0.0046
30				0.0004	31			0.0023
					33			0.0011
					35			0.0005

Statistics

Arithmetic mean

For a set of n observations, $x_1, x_2, x_3 \ldots x_n$

$$\bar{x} = \frac{\sum_{i=1}^{n} x_i}{n}$$

For a frequency distribution

$$\bar{x} = \frac{\sum_{i=1}^{n} f_i x_i}{\sum_{i=1}^{n} f_i}.$$

Median

For a series of observations arranged in ascending (or descending) order of size the median is the middle value. If the number of observations is even the median is the mean of the two middle values.

For a frequency distribution

$$\text{median} = L + i\left(\frac{\frac{1}{2}N - c}{f}\right)$$

L = lower boundary of the median class
i = width of median class
N = total frequency
c = cumulative frequency up to median class
f = frequency of the median class

Mode

For a set of observations the mode is the value which occurs most frequently.

For a frequency distribution

$$\text{mode} = L + i\left(\frac{d_1}{d_1 + d_2}\right)$$

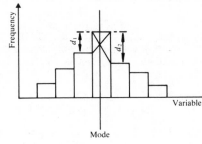

L = lower boundary of the modal class
i = width of modal class
d_1 = difference between frequency of modal class and frequency of next lower class
d_2 = difference between frequency of modal class and frequency of next higher class
w_i = weighting factors

Weighted arithmetic mean

$$\bar{x} = \frac{\sum_{i=1}^{n} w_i x_i}{\sum_{i=1}^{n} w_i}$$

Empirical relationship between mean, median and mode

For moderately skewed distributions which are unimodal

$$\text{Mean} - \text{mode} = 3(\text{mean} - \text{median})$$

Geometric mean For n values $x_1, x_2, x_3, \ldots, x_n$ $\qquad G = \sqrt[n]{x_1, x_2, x_3, \ldots, x_n}$.

For a frequency distribution $G = \sqrt[N]{x_1^{f_1}, x_2^{f_2}, x_3^{f_3}, \ldots, x_n^{f_n}}$

Harmonic mean This is the reciprocal of the arithmetic mean $\qquad H = \dfrac{n}{n\sum\limits_{i=1}^{n} \dfrac{1}{x}}$.

Relation between geometric mean, harmonic mean and arithmetic mean

$$H \leqslant G \leqslant \bar{x}$$

The three means have equal values only if all the observations in the set are the same.

Root mean square

$$\text{r.m.s.} = \sqrt{\frac{\sum\limits_{i=1}^{n} x_i^2}{n}}$$

Quartiles The quartiles are the values which divide the set of observations into four equal parts. Q_1, Q_2, Q_3 are the 1st, 2nd and 3rd quartiles respectively, Q_2 being equal to the median.

Deciles The values which divide the data into ten equal parts are called deciles. They are denoted by $D_1, D_2, \ldots D_9$.

Percentiles These are the values which divide the data into one-hundred equal parts. They are denoted by $P_1, P_2 \ldots P_{99}$. The 5th decile and the 50th percentile correspond to the median and the 25th and 75th percentiles correspond to the 1st and 3rd quartiles respectively.

The range Range = largest observation − smallest observation.

Mean deviation For a set of observations, $x_1, x_2 \ldots x_n$

$$\text{M.D.} = \frac{\sum\limits_{i=1}^{n} |x_i - \bar{x}|}{n} \qquad \bar{x} = \text{arithmetic mean}$$

For a frequency distribution

$$\text{M.D.} = \frac{\sum\limits_{i=1}^{n} f_i|x_i - \bar{x}|}{n}$$

Semi-interquartile range $Q = \dfrac{Q_3 - Q_1}{2}$ $\qquad$ Q_3 = 3rd quartile (or upper quartile)
Q_1 = 1st quartile (or lower quartile)

Standard deviation

For a set of observations, $x_1, x_2 \ldots x_n$

$$\sigma = \sqrt{\frac{\sum_{i=1}^{n} x_i^2}{n} - \bar{x}^2}$$ $\bar{x} = $ arithmetic mean

For a frequency distribution

$$\sigma = \sqrt{\frac{\sum_{i=1}^{n} fx^2}{\sum_{i=1}^{n} f_i} - \bar{x}^2}$$

Variance

$\text{Var}(x) = \sigma^2$.

Sheppard's correction for variance

To adjust for error in grouping data into classes

corrected variance $= \text{var}(x) - \dfrac{c^2}{12}$ $\text{var}(x) = $ variance from grouped data
$c = $ class width

Empirical relationships between measures of dispersion

For moderately skewed distributions:

mean deviation $= \frac{4}{5}$ standard deviation
semi-interquartile range $= \frac{2}{3}$ standard deviation

Coefficient of variation

$$V = \frac{\sigma}{\bar{x}} \times 100$$ $\sigma = $ standard deviation
$\bar{x} = $ arithmetic mean

Moments

For a set of observations, $x_1, x_2 \ldots x_n$

$$\bar{x}^r = \frac{\sum_{i=1}^{n} x^r}{n} \quad = \quad \bar{x}^r = \text{the } r\text{th moment}$$

(The first moment, i.e. $r = 1$, is equal to the arithmetic mean.)

The rth moment about any origin A is

$$m_r' = \frac{\sum_{i=1}^{n} (x - A)^r}{n}$$

The rth moment about the arithmetic mean is

$$m_r = \frac{\sum_{i=1}^{n} (x - \bar{x})^r}{n}$$

For a frequency distribution

$$\bar{x}^r = \frac{\sum\limits_{i=1}^{n} f_i x_i^r}{\sum\limits_{i=1}^{n} f_i}$$

$$m_r = \frac{\sum\limits_{i=1}^{n} f_i (x_i - \bar{x})^r}{\sum\limits_{i=1}^{n} f_i} \qquad \text{(Note that } m_2 = \sigma^2\text{)}$$

$$m_r' = \frac{\sum\limits_{i=1}^{n} f_i (x_i - A)^r}{\sum\limits_{i=1}^{n} f_i}$$

Skewness

Pearson's 1st coefficient of skewness $= \dfrac{3(\bar{x} - \text{mode})}{\sigma}$

Pearson's 2nd coefficient of skewness $= \dfrac{3(\bar{x} - \text{median})}{\sigma}$

Quartile coefficient of skewness $= \dfrac{(Q_3 - Q_2) - (Q_2 - Q_1)}{Q_3 - Q_1}$

Moment coefficient of skewness $= \dfrac{m_3}{\sigma^3} = \dfrac{m_3}{m_2^{3/2}}$

Kurtosis

This is a degree of peakedness of a distribution, taken relative to the normal distribution.

$$b_2 = \frac{m_4}{\sigma^4} = \frac{m_4}{m_2^2}$$

$b_2 =$ moment coefficient of kurtosis
$\sigma =$ standard deviation
$m_i = i$th moment

For a normal distribution $b_2 = 3$. A leptokurtic distribution has $(b_2 - 3)$ positive whilst $(b_2 - 3)$ is negative for a platykurtic distribution.

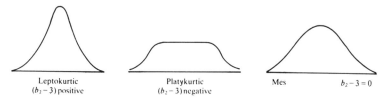

Leptokurtic	Platykurtic	Mes
$(b_2 - 3)$ positive	$(b_2 - 3)$ negative	$b_2 - 3 = 0$

Percentile coefficient of kurtosis $= \dfrac{\frac{1}{2}(Q_3 - Q_1)}{P_{90} - P_{10}}$

(For the normal distribution the percentile coefficient of kurtosis $= 0.263$.)

Binomial distribution	If p and $q = p - 1$ are the probabilities of success and failure respectively in a singl trial then the probabilities of $0, 1, 2, \ldots$ successes in n trials are given by th successive terms of the expansion of $(q + p)^n$.

Number of successes	0	1	2	3
Probability	q^n	$nq^{n-1}p$	$\dfrac{n(n-1)}{2!}q^{n-2}p^2$	$\dfrac{n(n-1)(n-2)}{3!}q^{n-3}p^3$

The normal distribution

The normal curve is defined by the equation

$$y = \frac{1}{\sigma\sqrt{2\pi}}\exp\left(-\tfrac{1}{2}(x+\bar{x})^2/\sigma^2\right)$$

To use the table of areas under the normal curve use the transformation

$$u = \frac{\dot{x}-\bar{x}}{\sigma}$$

Poisson distribution

$$p(x) = \frac{\lambda^x e^{-\lambda}}{x!}$$

λ is the average number of random events in a given interval. If the probability c success in a single trial is p, then in n trials $\lambda = np$.

Number of successes	0	1	2	3
Probability	$e^{-\lambda}$	$\lambda e^{-\lambda}$	$\dfrac{\lambda^2 e^{-\lambda}}{2!}$	$\dfrac{\lambda^3 e^{-\lambda}}{3!}$

Relation between the binomial, poisson and normal distributions

If $n \geqslant 50$ and $\lambda = np < 5$ then the Binomial distribution is very closely approx mated by the Poisson distribution.

If $n = 50$ and $\lambda = np > 5$ then the Binomial distribution is well approximated by th Normal distribution with $\bar{x} = np$ and $\sigma = \sqrt{npq}$.

Sampling distributions

$$\sigma_{\bar{x}} = \frac{\sigma}{\sqrt{n}}$$

$$\mu_{\bar{x}} = \mu$$

$\sigma_{\bar{x}}$ = standard deviation for sample means
σ = standard deviation for distribution of individual item
n = number of items in the sample
$\mu_{\bar{x}}$ = mean of sample means
μ = mean of individual items

Confidence levels

95% confidence limit $= \bar{x} \pm 1.96\sigma_{\bar{x}}$ $\bar{x}$ = sample mean
99% confidence limit $= \bar{x} \pm 2.58\sigma_{\bar{x}}$

To test if a sample having a mean $\bar{x}$ could have been drawn from a population with a mean μ,

$$t = \frac{|\mu - \bar{x}|}{\sigma/\sqrt{n}}$$

σ = population s.d. s = sample s.d.
μ = mean of population.
n = number in the sample.

If $n > 30$ then $\sigma = s$ and the normal distribution is used.

If $n \leqslant 30$ then $\sigma = s\sqrt{\dfrac{n}{n-1}}$ and the t-test is used.

In both cases $\mu = \mu_{\bar{x}}$.

To test for differences of means:

$$t = \frac{|\bar{x}_1 - \bar{x}_2|}{\sigma_d}$$

$\bar{x}_1$ = mean of sample 1
$\bar{x}_2$ = mean of sample 2

$$\sigma_d = \sqrt{\frac{s_1^2}{n_1} + \frac{s_2^2}{n_2}}$$

s_1 = s.d. of sample 1
s_2 = s.d. of sample 2
n_1 = number of items in sample 1
n_2 = number of items in sample 2

$$\chi^2 = \sum_{i=1}^{n} \frac{(O_i - E_i)^2}{E_i}$$

O = observed frequencies
E = estimated frequencies

If s_A = standard deviation of sample A

s_B = standard deviation of sample B

n_A = number in sample A

n_B = number in sample B

Then $\sigma_A = s_A\sqrt{\dfrac{n_A}{n_A - 1}}$ and $\sigma_B = s_B\sqrt{\dfrac{n_B}{n_B - 1}}$

$$F = \frac{\text{greatest } \sigma^2}{\text{least } \sigma^2}$$

With degrees of freedom for sample $A = n_A - 1$ and degrees of freedom for sample $B = n_B - 1$.

The least square line approximating the set of points (x_1, y_1), $(x_2, y_2) \ldots (x_n, y_n)$ has the equation

$$y = a_0 + a_1 x$$

The constants a_0 and a_1 are found by solving the equations

$$\sum y = a_0 n + a_1 \sum x \tag{1}$$
$$\sum xy = a_0 \sum x + a_1 \sum x^2 \tag{2}$$

where n is the number of points in the set.

The amount of work involved in finding the least square line can be shortened by using

$$y = \left(\frac{\sum XY}{\sum X^2}\right) x$$

where $X = x - \bar{x}$ and $Y = y - \bar{y}$.

Least square parabola

The least square parabola approximating to the set of points (x_1, y_1), $(x_2, y_2) \ldots (x_n, y_n)$ has the equation

$$y = a_0 + a_1 x + a_2 x^2$$

The constants a_0, a_1 and a_2 are found by solving the equations

$$\sum y = a_0 n + a_1 \sum x + a_2 \sum x^2 \tag{1}$$

$$\sum xy = a_0 \sum x + a_1 \sum x^2 + a_2 \sum x^3 \tag{2}$$

$$\sum x^2 y = a_0 \sum x^2 + a_1 \sum x^3 + a_2 \sum x^4 \tag{3}$$

Coefficient of correlation

$$r = \sqrt{\frac{\sum (x - \bar{x})(y - \bar{y})}{\sum x^2 \sum y^2}} = \frac{\sum xy - n\bar{x}\bar{y}}{n\sigma_x \sigma_y}$$

For perfect positive correlation $r = +1$ and for perfect negative correlation $r = -1$.

Rank correlation

Spearman's formula for rank correlation is:

$$r_S = 1 - \frac{6 \sum D^2}{n(n^2 - 1)}$$

D = difference in ranks of the corresponding values of x and y

n = the number of pairs of values of x and y in the set

Kendall's rank correlation coefficient is given by:

$$\tau = \frac{4N+}{n(n-1)} - 1$$

$N+$ = the number of pairs in the second ranking that are in the correct order assuming that the first ranking is in the correct order

Moving averages

For the set of numbers $x_1, x_2, x_3 \ldots$ the moving average of order n is given by the following sequence of arithmetic means:

$$\frac{x_1 + x_2 + \ldots + x_n}{n}; \quad \frac{x_2 + x_3 + \ldots + x_{n+1}}{n}; \quad \frac{x_3 + x_4 + \ldots + x_{n+2}}{n}$$

Index numbers

price relative $= \dfrac{p_n}{p_0}$

quantity or volume relative $= \dfrac{q_n}{q_0}$

value relative $= \dfrac{p_n q_n}{p_0 q_0}$

p_0 = price during the base period

p_n = price during the given period

q_n = quantity or volume during the given period

q_0 = quantity or volume during the base period

Laspeyres' index $= \dfrac{\sum p_n q_0}{\sum p_0 q_0}$

Paasche's index $= \dfrac{\sum p_n q_n}{\sum p_0 q_n}$

Fisher's ideal index $= \sqrt{\dfrac{\sum p_n q_0}{\sum p_0 q_0} \times \dfrac{\sum p_n q_n}{\sum p_0 q_n}}$

(Fisher's ideal index is the geometric mean of Laspeyres' index and Paasche's index.)

Marshall–Edgeworth price index $= \dfrac{\sum p_n(q_0 + q_n)}{\sum p_0(q_0(q_0 + q_n)}$

Laspeyres' volume index $= \dfrac{\sum q_n p_0}{\sum q_0 p_0}$.

Applied mathematics

Composition of vectors

If the angle between two vectors is θ and their magnitudes are a and b, the magnitude of the resultant is:

$$R = \sqrt{a^2 + b^2 + 2ab \cos \theta}$$

Triangle of forces

If three forces acting at a point are in equilibrium, the vectors representing the forces form, when added, a closed triangle. In order to be in equilibrium, three non-parallel forces must be concurrent.

Polygon of forces

If four or more forces acting at a point are in equilibrium, the vectors representing the forces form, when added, a closed polygon.

Principle of moments

For equilibrium, the sum of the clockwise moments is equal to the sum of the anti-clockwise moments about any point.

Moment

$M = Fd$

M = moment of force (Nm)
F = force (N)
d = distance measured perpendicular to the line of action of the force (m)

Conditions of equilibrium

The sum of the horizontal forces must be zero $\quad \sum F_H = 0$

The sum of the vertical forces must be zero $\quad \sum F_V = 0$

The sum of the moments must be zero $\quad \sum M = 0$

Lami's theorem

If the three forces P, Q and R are in equilibrium then

$$\frac{P}{\sin \alpha} = \frac{Q}{\sin \beta} = \frac{R}{\sin \gamma}$$

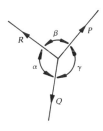

Friction

$$\mu = \frac{F}{N} = \tan \lambda$$

μ = coefficient of friction (no units)
F = force required to move one surface over another (N)
N = force pressing the surfaces together (N)
λ = the angle of friction.

At all times $F \leqslant \mu N$.

Centre of gravity

$$\bar{x} = \frac{\sum wx}{\sum w} \qquad \bar{y} = \frac{\sum xy}{\sum w}$$

Uniform velocity

If s metres is the distance travelled in a time t seconds the velocity is $v = \dfrac{s}{t}$.

Uniformly accelerated motion

$$v_f = v_i + at$$
$$s = v_i t + \tfrac{1}{2}at^2$$
$$s = \left(\frac{v_i + v_f}{2}\right)t$$
$$v_f^2 = v_i^2 + 2as$$

v_f = final velocity (m s^{-1})
v_i = initial velocity (m s^{-1})
s = distance travelled (m)
t = time (s)
a = acceleration (m s^{-2})

Angular motion

$$\omega_t = \omega_i + \alpha t$$
$$\theta = \omega_i t + \tfrac{1}{2}\alpha t^2$$
$$\theta = \left(\frac{\omega_i + \omega_2}{2}\right)t$$
$$\omega_f^2 = \omega_i^2 + 2\alpha s$$
$$\omega = \frac{2\pi N}{60}$$

ω_f = final angular velocity (rad s^{-1})
ω_i = initial angular velocity (rad s^{-1})
θ = angle turned (rad)
t = time (s)
α = angular acceleration (rad s^{-2})
N = angular speed (rev/min)

Falling bodies

$$v = gt$$
$$s = \tfrac{1}{2}gt^2$$
$$v = \sqrt{2gh}$$

v = velocity attained after t seconds (m s^{-}
g = acceleration due to gravity (m s^{-2})
s = distance fallen after t seconds (m)
h = height fallen in t seconds (m)

Bodies projected upwards

$$t = \frac{v}{g}$$
$$h_{max} = \frac{v^2}{2g}$$

t = time to reach greatest height (s)
v = velocity of projection (m s^{-1})
h_{max} = greatest height (m)
g = acceleration due to gravity (m s^{-2})

Projectiles

For a body projected with a velocity v metres per second at an angle α with the horizontal:

Time to highest point of flight is $\quad t = \dfrac{v \sin \alpha}{g}$ seconds

Total time of flight is $\quad 2t$

Maximum height attained is $\quad h = \dfrac{v^2 \sin^2 \alpha}{2g}$ metres

Horizontal range is $\quad R = \dfrac{v^2 r \sin 2\alpha}{g}$ metres

Projectiles on an incline plane

If the body is projected up the plane

$$R = \frac{2v^2 \sin \beta \cos (\beta + \alpha)}{g \cos^2 \alpha}$$

For a given value of v the range is a maximum when

$$\sin (2\beta + a) = 1 \quad \text{and it is} \quad R_{max} = \frac{v^2(1 - \sin \alpha)}{g \cos^2 \alpha}$$

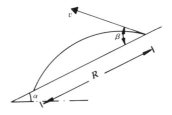

If the body is projected down the plane

$$R = \frac{2v^2 \sin \beta \cos (\beta - \alpha)}{g \cos^2 \alpha}$$

For a given value of v, R is a maximum when $\sin (2\beta - \alpha) = 1$ and it is

$$R_{max} = \frac{v^2(1 + \sin \alpha)}{g \cos^2 \alpha}$$

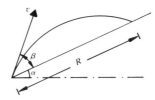

General motion of a particle in a plane

For motion in a straight line:

If $a = f(t)$, $\quad a = \dfrac{dv}{dt} = \dfrac{d^2 s}{dt^2}, \quad v = \dfrac{ds}{dt}$

If $a = f(t)$ $\qquad \displaystyle\int dv = \int f(t)\, dt$

If $a = f(v)$ $\qquad \displaystyle\int \frac{dv}{f(v)} = \int dt \quad \text{or} \quad \int \frac{v\, dv}{f(v)} = \int ds$

If $a = f(s)$ $\qquad \displaystyle\int v\, dv = \int f(s)\, ds$

where s = distance, t = time, v = velocity and a = acceleration.

The impulse of a variable force F is

$$\int_0^t F\, dt = mv_f - mv_i$$

The work done by a variable force F is

$$\int_0^s F\,\mathrm{d}s = \tfrac{1}{2}mv_f^2 - \tfrac{1}{2}mv_i^2$$

where m = mass, v_i = initial velocity and v_f = final velocity.

Work done

$W = Fs$

$W = T\theta$

T = torque (Nm)
θ = angle turned (rad)
W = work done (J)
F = force moving the body (N)
s = distance moved along the line of action of the force (m)

The work done against gravity is

$W = mgh$

m = mass being lifted (kg)
g = acceleration due to gravity (m s^{-2})
h = height lifted (m)

Power

$$P = \frac{\text{work done}}{\text{time taken}}$$

$$= \frac{Fs}{t} = Fv$$

$$= \frac{T\theta}{t} = 2\pi NT$$

P = power (W)
t = time taken (s)
v = velocity of the body (m s^{-1})
N = rotational speed (rev s^{-1})

Energy

Potential energy = mgh.   Kinetic energy = $\tfrac{1}{2}mv^2$.

Newton's laws of motion

Law 1 Every body will remain at rest or continue to move with uniform velocity unless an external force is applied to it.

Law 2 When an external force is applied to a body of constant mass the force produces an acceleration which is directly proportional to the force.

Law 3 Action and reaction are always equal and opposite.

Force

If a mass m increases to $m + \delta m$ whilst its speed increases from v to $v + \delta v$ in a time δt then the resultant force acting on the mass is

$$F = m\frac{\mathrm{d}v}{\mathrm{d}t} + v\frac{\mathrm{d}m}{\mathrm{d}t} \qquad (1)$$

If δm has a velocity of u before joining the mass m then

$$F = m\frac{\mathrm{d}v}{\mathrm{d}t} + (v - u)\frac{\mathrm{d}m}{\mathrm{d}t} \qquad (2)$$

If the mass of a body is constant then $\dfrac{\mathrm{d}m}{\mathrm{d}t} = 0$ and equations (1) and (2) reduce to

$F = ma$

F = force causing acceleration (N)
m = mass being accelerated (kg)
a = acceleration (m s^{-2})

Weight	$W = mg$	W = weight of the body (N)
		m = mass of the body (kg)
		g = acceleration due to gravity (m s^{-2})

Simple Harmonic Motion (S.H.M.)

A particle moves with S.H.M. if its acceleration is proportional to the displacement from a fixed point or fixed line and the acceleration is always directed towards the fixed point or line.

If a particle P is describing circular motion with constant angular velocity ω then, if the distance $OQ = x$, Q has acceleration

$$\ddot{x} = -\omega^2 x = -\omega^2 a \cos \omega t$$

The velocity of Q is

$$\dot{x} = \omega\sqrt{a^2 - x^2}$$

and $x = a \cos \omega t$ where a is the amplitude.

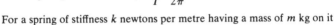

The periodic time is $T = \dfrac{2\pi}{\omega} = 2\pi\sqrt{\dfrac{x}{a}}$.

The frequency in hertz is $f = \dfrac{1}{T} = \dfrac{\omega}{2\pi}$.

For a spring of stiffness k newtons per metre having a mass of m kg on it

$$f = \frac{1}{2\pi}\sqrt{\frac{k}{m}}.$$

If the deflection in metres under the mass is δ_s then $\delta_s = \dfrac{mg}{k}$ and

$$f = \frac{1}{2\pi}\sqrt{\frac{g}{\delta_s}}.$$

Simple pendulum

The periodic time is $T = 2\pi\sqrt{\dfrac{l}{g}}$ where l is the length of the pendulum.

Damped harmonic oscillations

When a particle of mass m moves along a straight line Ox under the action of a force $m\omega^2 x$ directed towards O and a resisting force $2mkv$, the equation of motion is

$$\frac{d^2 x}{dt^2} + 2k\frac{dx}{dt} + \omega^2 x = 0$$

Provided that $k < \omega$, the solution of this equation is

$$x = C\,e^{-kt} \cos\left(t\sqrt{\omega^2 - k^2} + \alpha\right)$$

where C and α are constants of integration. e^{-kt} is called the damping factor and the amplitude of the oscillations decreases with time in geometric progression.

The period of the oscillations is constant and is given by $T = \dfrac{2\pi}{\sqrt{\omega^2 - k^2}}$.

If $k \geqslant n$ the motion is not oscillatory.

107

Hooke's Law

Elastic strings and springs obey Hooke's Law

$$T = \lambda \frac{x}{a}$$

(λ is equal to the force required to double the length of the string or spring)

T = tension in the string (N)
x = extension (or compression)
a = natural length
λ = modulus of elasticity (N)

The work done in stretching an elastic string is $W = \dfrac{\lambda x^2}{2a}$.

(The work done in compressing a spring an amount x is given by the same expression.)

Momentum

$$M = mv$$

M = momentum (kg m s^{-1})
m = mass (kg)
v = velocity (m s^{-1})

If a mass of m kilograms has its velocity changed from v_i metres per second to v_f metres per second by the action of a force F newtons acting for t seconds, then

$$Ft = m(v_f - v_i)$$

The quantity Ft is called the impulse.

Impact

$$m_A v_A + m_B v_B = m_A v'_A + m_B v'_B$$

$$e = \frac{v'_A - v'_B}{v_A - v_B}$$

v = velocity before impact (m s^{-1})
v' = velocity after impact (m s^{-1})
e = coefficient of restitution

Torque

$$T = Fr = I\alpha$$

$$I = mk^2$$

If a torque T acts on a body for a time t causing the angular velocity to change from ω_1 to ω_2 then

$$Tt = I(\omega_2 - \omega_1)$$

T = torque (Nm)
F = force (N)
r = radius at which the force is applied (m)
I = momentum of inertia (kg m^2)
α = angular acceleration (rad s^{-2})
m = mass (kg)
k = radius of gyration (m)

Uniform circular motion

$$\omega = \frac{v}{r} = \frac{2\pi}{T}$$

$$a = \frac{v^2}{r} = \omega^2 r = \frac{4\pi^2 r}{T^2}$$

$$F = \frac{mv^2}{r} = m\omega^2 r$$

ω = angular velocity (rad s^{-1})
v = tangential velocity (m s^{-1})
r = radius of the circle (m)
T = periodic time = time for 1 revolution (s)
a = acceleration towards the centre = centripetal acceleration (m s^{-2})
F = centrifugal force (N)
m = mass (kg)
T = tension in string (N)
l = length of pendulum (m)

Conical pendulum

$$T = ml\omega^2$$

Motion in a vertical circle

If a mass m moves in a vertical circle of radius a and passes the lowest point of the circle with speed u, then as shown in the diagram

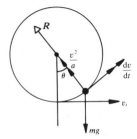

$$R - mg\cos\theta = m\frac{v^2}{a} \qquad \text{radially}$$

$$mg\sin\theta = -m\frac{dv}{dt} \qquad \text{tangentially}$$

$$\tfrac{1}{2}mv_i^2 - mga = \tfrac{1}{2}mv_f^2 - mga\cos\theta$$

A mass restricted to a circular path will describe complete circles if $v_i^2 > 4gr$ otherwise it will come momentarily to rest before reaching the highest point of the circle and subsequently it will oscillate. If the motion is not restricted to a circular path, as in the case of a particle rotating at one end of a light string, then

$$T = m\left(\frac{v_i^2}{a} - 2g + 3g\cos\theta\right)$$

where T is the tension in the string.

If the motion is a complete circle the string is taut in the highest position and $u^2 \geqslant 5ga$.

For oscillations $u^2 \leqslant 2ga$.

The range of values for which the string goes slack is $\sqrt{2ga} < u < \sqrt{5ga}$.

Vectors

Points and distances

If $\mathbf{r} = a\mathbf{i} + b\mathbf{j} + c\mathbf{k}$ is a vector and $\mathbf{i}$, $\mathbf{j}$ and $\mathbf{k}$ are unit vectors in the directions Ox, Oy and Oz the magnitude or modulus of $\mathbf{r}$ is

$$|\mathbf{r}| = \sqrt{a^2 + b^2 + c^2}$$

The direction cosines of $\mathbf{r}$ are

$$l = \frac{a}{|\mathbf{r}|}, \quad m = \frac{b}{|\mathbf{r}|} \quad \text{and} \quad n = \frac{c}{|\mathbf{r}|}$$

and $l^2 + m^2 + n^2 = 1$. Also

$$l : m : n = a : b : c.$$

If $\hat{\mathbf{r}}$ is a unit vector in the direction of $\mathbf{r}$

$$\hat{\mathbf{r}} = l\mathbf{i} + m\mathbf{j} + n\mathbf{k}$$

$$\mathbf{r} = r(l\mathbf{i} + m\mathbf{j} + n\mathbf{k}) \text{ where } r = |\mathbf{r}|$$

The position vector of the point $P(x, y, z)$ is

$$\mathbf{r} = x\mathbf{i} + y\mathbf{j} + z\mathbf{k} = \overrightarrow{OP}$$

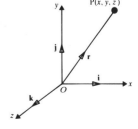

If two points A and B have coordinates x_1, y_1, z_1 and x_2, y_2, z_2 then

$$\overrightarrow{AB} = (x_2\mathbf{i}+y_2\mathbf{j}+z_2\mathbf{k})-(x_1\mathbf{i}+y_1\mathbf{j}+z_1\mathbf{k})$$

$$AB=|\overrightarrow{AB}| = \sqrt{(x_2-x_1)^2+(y_2-y_1)^2+(z_2-z_1)^2}$$

If P divides AB in the ratio $m:n$

$$\overrightarrow{OP} = \frac{n\mathbf{a}-m\mathbf{b}}{m+n}$$

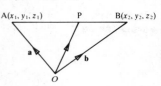

P has the coordinates:

$$\frac{nx_1+mx_2}{m+n}, \frac{ny_1+my_2}{m+n}, \frac{nz_1+mz_2}{m+n}$$

Products and angles between vectors

If θ is the angle between two vectors $\mathbf{a}$ and $\mathbf{b}$

where $\mathbf{a} = a_1\mathbf{i}+a_2\mathbf{j}+a_3\mathbf{k}=a(l\mathbf{i}+m\mathbf{j}+n\mathbf{k})$

and $\mathbf{b} = b_1\mathbf{i}+b_2\mathbf{j}+b_3\mathbf{k}=b(l'\mathbf{i}+m'\mathbf{j}+n'\mathbf{k})$

then the *scalar (or dot) product* of $\mathbf{a}$ and $\mathbf{b}$ is

$$\mathbf{a}\,.\,\mathbf{b} = ab\cos\theta = a_1b_1+a_2b_2+a_3b_3 = \mathbf{b}\,.\,\mathbf{a}$$

$$\cos\theta = \frac{\mathbf{a}\,.\,\mathbf{b}}{ab} = \hat{\mathbf{a}}\,.\,\hat{\mathbf{b}} = ll'+mm'+nn'$$

The *vector product* of $\mathbf{a}$ and $\mathbf{b}$ is

$$\mathbf{a}\times\mathbf{b} = \begin{vmatrix} \mathbf{i} & \mathbf{j} & \mathbf{k} \\ a_1 & a_2 & a_3 \\ b_1 & b_2 & b_3 \end{vmatrix} \begin{aligned} &= (a_2b_3-a_3b_2)\mathbf{i}-(a_1b_3-a_3b_1)\mathbf{j} \\ &\quad +(a_1b_2-a_2b_1)\mathbf{k} \\ &= ab\sin\theta\,\hat{\mathbf{n}} \end{aligned}$$

where $\hat{\mathbf{n}}$ is a unit vector perpendicular to $\mathbf{a}$ and $\mathbf{b}$ in the direction of a right-handed screw turned from $\mathbf{a}$ to $\mathbf{b}$.

$\mathbf{a}$ and $\mathbf{b}$ are perpendicular if $\mathbf{a}\,.\,\mathbf{b}=0$.

$\mathbf{a}$ and $\mathbf{b}$ are parallel if $\mathbf{a}\times\mathbf{b}=0$.

If $\mathbf{a}, \mathbf{b}$ and $\mathbf{c}$ are three vectors, the scalar quantity

$$V = \mathbf{a}\,.\,\mathbf{b}\times\mathbf{c} = \mathbf{b}\,.\,\mathbf{c}\times\mathbf{a} = \mathbf{c}\,.\,\mathbf{a}\times\mathbf{b}$$

is called the *triple scalar product*. V is the volume of a parallelepiped with sides $\mathbf{a}, \mathbf{b}$ and $\mathbf{c}$. If $V=0$ the vectors are coplanar.

If

$$\mathbf{a} = a_1\mathbf{i}+a_2\mathbf{j}+a_3\mathbf{k}, \quad \mathbf{b}=b_1\mathbf{i}+b_2\mathbf{j}+b_3\mathbf{k}$$

and

$$\mathbf{c}=c_1\mathbf{i}+c_2\mathbf{j}+c_3\mathbf{k}$$

then

$$\mathbf{a}\,.\,(\mathbf{b}\times\mathbf{c}) = \begin{vmatrix} a_1 & a_2 & a_3 \\ b_1 & b_2 & b_3 \\ c_1 & c_2 & c_3 \end{vmatrix}$$

The vector quantity

$$\mathbf{a}\times(\mathbf{b}\times\mathbf{c}) = (\mathbf{a}\,.\,\mathbf{c})\mathbf{b}-(\mathbf{a}\,.\,\mathbf{b})\mathbf{c}$$

is called a *triple vector product*. In general

$$\mathbf{a}\times(\mathbf{b}\times\mathbf{c}) \neq (\mathbf{a}\times\mathbf{b})\times\mathbf{c}$$

Straight lines　　If a straight line is parallel to a vector **b** and passes through the point $A(x_1, y_1, z_1)$ then the vector equation of the line is

$$\mathbf{r} = \mathbf{a} + \lambda \mathbf{b}$$

where　　**r** is the position vector of any point on the line

and　　　**a** is the position vector of the point A

If　　$\mathbf{b} = A\mathbf{i} + B\mathbf{j} + C\mathbf{k}$

$$\mathbf{r} = x_1\mathbf{i} + y_1\mathbf{j} + z_1\mathbf{k} + \lambda(A\mathbf{i} + B\mathbf{j} + C\mathbf{k})$$

where　　$\lambda = \dfrac{x - x_1}{A} = \dfrac{y - y_1}{B} = \dfrac{z - z_1}{C}$.

If l, m and n are the direction cosines of the line

$$l : m : n = A : B : C$$

If θ is the angle between two lines with vector equations

$$\mathbf{r} = \mathbf{a}_1 + \lambda \mathbf{b}_1 \quad \text{and} \quad \mathbf{r} = \mathbf{a}_2 + \lambda \mathbf{b}_2$$

then　　$\cos \theta = \dfrac{\mathbf{b}_1 . \mathbf{b}_2}{b_1 b_2}$.

If the point C has position vector **c** then its distance from the line whose vector equation is $\mathbf{r} = \mathbf{a} + \lambda \mathbf{b}$ is

$$d = \frac{1}{b}\left|(\mathbf{c} - \mathbf{a}) \times \mathbf{b}\right|$$

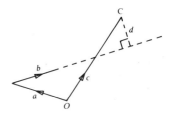

If two skew lines have vector equations

$$\mathbf{r} = \mathbf{a}_1 + \lambda \mathbf{b}_1$$

and

$$\mathbf{r} = \mathbf{a}_2 + \lambda \mathbf{b}_2$$

then the shortest distance between the lines is

$$\left|\frac{(\mathbf{a}_1 - \mathbf{a}_2) . (\mathbf{b}_1 \times \mathbf{b}_2)}{|\mathbf{b}_1 \times \mathbf{b}_2|}\right|$$

The lines intersect if　$(\mathbf{a}_1 - \mathbf{a}_2) . (\mathbf{b}_1 \times \mathbf{b}_2) = 0$.

Curves

Curve	Cartesian equation	Parametric equations	Vector equation
Circle	$x^2 + y^2 = a^2$	$x = a \cos \theta, \quad y = a \sin \theta$	$\mathbf{r} = a \cos \theta \mathbf{i} + a \sin \theta \mathbf{j}$
Parabola	$y^2 = 4ax$	$x = at^2, \quad y = 2at$	$\mathbf{r} = at^2\mathbf{r} + 2at\mathbf{j}$
Ellipse	$\dfrac{x^2}{a^2} + \dfrac{y^2}{b^2} = 1$	$x = a \cos \theta, \quad y = b \sin \theta$	$\mathbf{r} = a \cos \theta \mathbf{i} + b \sin \theta \mathbf{j}$
Hyperbola	$\dfrac{x^2}{a^2} - \dfrac{y^2}{b^2} = 1$	$x = a \sec \theta, \quad y = b \tan \theta$	$\mathbf{r} = a \sec \theta \mathbf{i} + b \tan \theta \mathbf{j}$
Rectangular hyperbola	$xy = c^2$	$x = ct, \quad y = \dfrac{c}{t}$	$\mathbf{r} = ct\mathbf{i} + \dfrac{c}{t}\mathbf{j}$
Helix			$\mathbf{r} = a \cos \theta \mathbf{i} + a \sin \theta \mathbf{j} + \dfrac{p\theta}{2\pi}\mathbf{k}$

Planes

If $\mathbf{n}$ is a vector perpendicular to a plane and $\dfrac{D}{|\mathbf{n}|} = d$ is the distance of the plane from the origin then the *vector equation of the plane* can be written as

$$\mathbf{r} \cdot \mathbf{n} = D \quad \text{or} \quad \mathbf{r} \cdot \hat{\mathbf{n}} = d$$

If $\mathbf{a}$ is the position vector of a point on a plane and $\mathbf{p}$ and $\mathbf{q}$ are parallel to the plane, the *parametric form* of the vector equation of the plane is

$$\mathbf{r} = \mathbf{a} + \lambda \mathbf{p} + \mu \mathbf{q}$$

If $\mathbf{a}$, $\mathbf{b}$ and $\mathbf{c}$ are the position vectors of non-colinear points in the plane then

$$\mathbf{r} = (1 - \lambda - \mu)\mathbf{a} + \lambda \mathbf{b} + \mu \mathbf{c}$$

If A, B and C are the direction ratios of the normal to the plane, the *Cartesian form* of the equation of the plane is

$$Ax + By + Cz = D$$

If $\mathbf{r} \cdot \hat{\mathbf{n}}_1 = d_1$ and $\mathbf{r} \cdot \hat{\mathbf{n}}_2 = d_2$ are the vector equations of two planes then the angle θ between the planes is given by

$$\cos \theta = \hat{\mathbf{n}}_1 \cdot \hat{\mathbf{n}}_2$$

The distance of a point P from the plane $\mathbf{r} \cdot \hat{\mathbf{n}} = d$ is given by $\mathbf{r} \cdot \mathbf{a} - d$, where $\mathbf{a}$ is the position vector of P.

Differentiation of a vector with respect to a scalar variable

If $\mathbf{r} = f(\theta)\mathbf{i} + g(\theta)\mathbf{j} + h(\theta)\mathbf{k}$

$$\frac{d\mathbf{r}}{d\theta} = f'(\theta)\mathbf{i} + g'(\theta)\mathbf{j} + h'(\theta)\mathbf{k}$$

$$\frac{d\mathbf{r}}{dt} = \frac{d\mathbf{r}}{d\theta}\frac{d\theta}{dt} = f'(\theta)\frac{d\theta}{dt}\mathbf{i} + g'(\theta)\frac{d\theta}{dt}\mathbf{j} + h'(\theta)\frac{d\theta}{dt}\mathbf{k}$$

where θ and t are scalar variables.

Differentiation of a unit vector with respect to a scalar

If $\hat{\mathbf{r}}$ is a unit vector such that

$$\hat{\mathbf{r}} = l\mathbf{i} + m\mathbf{j} + n\mathbf{k}$$

where l, m and n are the variable direction cosines then

$$\hat{\mathbf{r}}\frac{d\hat{\mathbf{r}}}{dt} = 0$$

t being a scalar variable.

Motion of a particle in a plane

If a particle P moves along a curve in the xy plane so that at time t, P is at the point with polar co-ordinates (r, θ), then the position vector of P at time t is

$$\mathbf{r} = r(\cos \theta\, \mathbf{i} + \sin \theta\, \mathbf{j})$$

$$\mathbf{v} = \frac{d\mathbf{r}}{dt} = \dot{r}\hat{\mathbf{r}} + r\dot{\theta}\hat{\mathbf{s}}$$

$$\mathbf{a} = \frac{d\mathbf{v}}{dt} = (\ddot{r} - r\dot{\theta}^2)\hat{\mathbf{r}} + (r\ddot{\theta} + 2\dot{r}\dot{\theta})\hat{\mathbf{s}}$$

where $\hat{\mathbf{s}} = -\sin \theta\, \mathbf{i} + \cos \theta\, \mathbf{j}$

The radial component of velocity is $\dot{r}$ and its transverse component is $r\dot{\theta}$.

The radial component of acceleration is $\ddot{r} - r\dot{\theta}$ and its transverse component is $r\ddot{\theta} + 2\dot{r}\dot{\theta}$.

Standard symbols and units for physical quantities

Quantity	Symbol	Unit	Quantity	Symbol	Unit
Acceleration–gravitational	g	m/s²	Frequency	f	Hz
Acceleration–linear	a	m/s²	Frequency, resonant	f_r	Hz
Admittance	Y	S	Gravitational acceleration	g	m/s²
Altitude above sea level	z	m	Gibbs function	G	J
Amount of substance	n	mol	Gibbs function, specific	g	kJ/kg
Angle–plane	$\alpha, \beta, \theta, \phi$	rad			
Angle–solid	Ω, ω	steradian	Heat capacity, specific	c	kJ/kg K
Angular acceleration	α	rad/s²	Heat flow rate	ϕ	W
Angular velocity	ω	rad/s	Heat flux intensity	ϕ	kW/m²
Area	A	m²			
Area–second moment of	I	m⁴	Illumination	E	lx
			Impedance	Z	Ω
Bulk modulus	K	N/m²	Inductance, self	L	H
			Inductance, mutual	M	H
Capacitance	C	μF	Internal energy	U, E	J
Capacity	V	l	Internal energy, specific	u, e	kJ/kg
Coefficient of friction	μ	no unit	Inertia, moment of	I, J	kg m²
Coefficient of linear					
expansion	α	/°C	Kinematic viscosity	ν	m²/s, St
Conductance, electrical	G	S			
Conductance, thermal	h	kW/m²K	Length	l	m
Conductivity, electrical	σ	kS/mm	Light–velocity of	c	m/s
Conductivity, thermal	λ	W/m K	Light–wavelength of	λ	m
Cubical expansion–			Linear expansion–		
coefficient of	β	/°C	coefficient of	α	/°C
Current, electrical	I	A	Luminance	L	cd/m²
Current density	J	A/mm²	Luminous flux	ϕ	lm
			Luminous intensity	I	cd
Density	ρ	kg/m³			
Density, relative	d	no unit	Magnetic field strength	H	A/m
Dryness fraction	x	no unit	Magnetic flux	Φ	Wb
Dynamic viscosity	η	Ns/m², cP	Magnetic flux density	B	T
			Magnetomotive force	F	A
Efficiency	η	no unit	Mass, macroscopic	m	kg
Elasticity, modulus of	E	N/m²	Mass, microscopic	M	u
Electric field strength	E	V/m	Mass, rate of flow	V	m³/s
Electric flux	ϕ	C	Mass, velocity	G	kg/m²s
Electric flux density	D	C/m²	Modulus, bulk	K	N/m²
Energy	W	J	Modulus of elasticity	E	N/m²
Energy, internal	U, E	J	Modulus of rigidity	G	N/m²
Energy, specific internal	u, e	kJ/kg	Modulus of section	Z	m³
Enthalpy	H	J	Molar mass of gas	M	kg/k mol
Enthalpy, specific	h	kJ/kg	Molar volume	V_0	m³/k mol
Entropy	S	kJ/K	Moment of force	M	Nm
Expansion–coefficient			Moment of inertia	I, J	kg m²
of cubical	β	/°C	Mutual inductance	M	H
Expansion–coefficient					
of linear	α	/°C	Number of turns in a		
			winding	N	no unit
Field strength, electric	E	V/m			
Field strength, magnetic	H	A/m	Periodic time	T	s
Flux density, electric	D	C/m²	Permeability, absolute	μ	μH/m
Flux density, magnetic	B	T	Permeability, absolute of		
Flux, electric	ψ	C	free space	μ_0	μH/m
Flux, magnetic	Φ	Wb	Permeability, relative	μ_r	
Force	F	N	Permeance	Λ	H
Force, resisting	R	N	Permittivity, absolute	ϵ	pF/m

Quantity	Symbol	Unit	Quantity	Symbol	Unit
Permittivity of free space	ϵ_0	pF/m	Stress, direct	σ	N/m²
Permittivity, relative	ϵ_r	no unit	Shear modulus of		
Poisson's ratio	ν	no unit	rigidity	G	N/m²
Polar moment of area	J	m⁴	Surface tension	γ	N/m
Power, apparent	S	VA	Susceptance	B	S
Power, active	P	W			
Power, reactive	Q	VA$_r$	Temperature value	θ	°C
Pressure	p	N/m²	Temperature coefficients		
			of resistance	α, β, γ	/°C
Quantity of heat	Q	J	Thermodynamic		
Quantity of electricity	Q	Ah, C	temperature value	T	K
			Time	t	s
Reactance	X	Ω	Torque	T	Nm
Reluctance	S	/H, A/Wb			
Relative density	d	no unit	Vapour velocity	C	m/s
Resistance, electrical	R	Ω	Velocity	v	m/s
Resisting force	R	N	Velocity, angular	ω	rad/s
Resistance, temperature			Velocity of light	c	Mm/s
coefficients of	α, β, γ	/°C	Velocity of sound	a	m/s
Resistivity, conductors	ρ	M Ω mm	Voltage	V	V
Resistivity, insulators	ρ	M Ω mm	Volume	V	m³
Resonant frequency	f_r	Hz	Volume, rate of flow	V	m³/s
			Viscosity, dynamic	η	Ns/m², cP
Second moment of area	I	m⁴	Viscosity, kinematic	ν	m²/s, cSt
Self inductance	L	H			
Shear strain	γ	no unit	Wavelength	λ	m
Shear stress	τ	N/m²	Work	W	J
Specific gas constant	R	kJ/kg K			
Specific heat capacity	c	kJ/kg K	Young's modulus of		
Specific volume	v	m³/kg	elasticity	E	N/m²
Strain, direct	ϵ	no unit			

Abbreviations for units

Unit	abb.	Unit	abb.	Unit	abb.	Unit	abb.
metre	m	steradian	sr	newton	N	mole	mol
angström	A	radian per		bar	bar	watt	W
square metre	m²	second	rad/s	millibar	mb	decibel	dB
cubic metre	m³	hertz	Hz	standard		kelvin	K
litre	l	revolution per		atmosphere	atm	centigrade	°C
second	s	minute	rev/min	millimetre of		coulomb	C
minute	min.	kilogramme	kg	mercury	mm Hg	ampere	A
hour	h	gramme	g	poise	P	volt	V
lumen	lm	tonne		stokes	S, St	ohm	Ω
candela	cd	(= 1 Mg)	t	joule	J	farad	F
lux	lx	seimen	S	kilowatt hour	kW h	henry	H
day	d	atomic mass		electron volt	eV	weber	Wb
year	a	unit	u	calorie	cal	tesla	T
radian	rad						

Index

Notes

Notes

Notes

$\frac{1}{x}$

$2\frac{7}{8}$

$3\frac{8}{16}\frac{2}{32} + 5\frac{9}{16}$

$9\frac{1}{16}$

$2\frac{7}{8} + 1\frac{3}{4} = 4.685$

$2.875 + 1.75 =$

$6\frac{5}{12} + 4\frac{9}{10}$

$45°, 36', 24''$

$\boxed{45.6067}$

$\frac{28}{8} \times \frac{14}{8} = \frac{37}{1}$